DATAM:
Digital Approaches to Teaching the Ancient Mediterranean

Edited by
Sebastian Heath

The Digital Press at the University of North Dakota
Grand Forks, ND

2020. The Digital Press at the University of North Dakota

Unless otherwise indicated, all contributions to this volume appear under a Creative Commons Attribution 4.0 License:
https://creativecommons.org/licenses/by/4.0/legalcode

Library of Congress Control Number: 2020935190

ISBN-13: 978-1-7345068-1-5 (ebook)
ISBN-13: 978-1-7345068-2-2 (Paperback)

Download a full color version of this book from:
http://thedigitalpress.org/datam/

DATAM:
Digital Approaches to Teaching
the Ancient Mediterranean

Table of Contents

Editor's Preface
   Sebastian Heath 1

Preface
   Helen Cullyer 5

Foreword
   Shawn Graham 9

Futures of Classics: Obsolescence and Digital Pedagogy
   Lisl Walsh 17

Teaching Information Literacy in the
Digital Ancient Mediterranean Classroom
   David M. Ratzan 31

Dissecting Digital Divides in Teaching
   William Caraher 71

Autodidacts and the "Promise" of Digital Classics
   Patrick J. Burns 83

Playing the Argonauts: Pedagogical Pathways through
Creation and Engagement in a Virtual Sea
   Sandra Blakely 97

Programming without Code: Teaching Classics and
Computational Methods
   Marie-Claire Beaulieu and Anthony Bucci 127

Digital Creation and Expression in the
Context of Teaching Roman Art and Archaeology
   Sebastian Heath 149

Digital Janiform: The Digital Object from
Research to Teaching
   Eric Poehler 171

Contributors 191

Editor's Preface

Sebastian Heath

An email that went out early in the process of organizing the conference *Digital Approaches to Teaching the Ancient Mediterranean* from which this volume follows, included the sentences:

> "Our focus is on what is working (and what isn't) in the undergraduate classroom. Actual assignments, syllabi, which cloud-based tools we use are all of interest. Stepping back and asking what new and '2018-relevant' topics can be brought into the classroom with digital approaches is also on the agenda. 'How and why?' are the questions when put most briefly."

That language itself worked, in part because no one took it to mean that there were strict bounds on the hoped for discussion. Accordingly, I write now with great gratitude to the speakers who came to NYU's Institute for the Study of the Ancient World (ISAW) in October of 2018 and shared their thoughts and practices with an active and engaged audience to whom I am also grateful. It was likewise a great pleasure to collaborate with my co-organizers at ISAW, Tom Elliott and David Ratzan, and to work with Helen Cullyer of the Society for Classical Studies (SCS), who was our partner in organizing the event. The idea for DATAM rose out of the series of digitally themed conferences that ISAW has been hosting since 2015. But it takes a lot of thinking and care to turn a vague idea - "How about digital tools and teaching?" - into an actual conference and much of both went into making DATAM a success. Long before the day itself, and then when ISAW was filled with participants, many colleagues at ISAW were essential contributors. I particularly thank Marc LeBlanc and Diane Bennett and their staffs for working on travel and lodging logistics and for helping to make ISAW a welcoming venue while DATAM was underway. ISAW offered financial support, as did the SCS, which made it possible to bring in speakers from distant campuses. Every contribution, of whatever form, was necessary and I repeat my gratitude for them all.

It was also early in the process that I reached out to William Caraher to ask if the University of North Dakota's Digital Press would be interested in publishing the papers. His initial interest and subsequent willingness to be a speaker were part of building early momentum. When it came to working with speakers to turn their presentations into the chapters that you will find in this volume, I also sent out emails. In one of those was the phrase, "[a] collection of thoughtful writing with a practical bent," that I used as a description of what the contributions could collectively be; though I was also clear that I was interested in each contributor writing in their own voice. I hoped that this language would come across as following naturally from the tenor of the event, and again, I believe it worked. All the essays combine both description of practice and the pedagogic thinking that informed that practice. I do take this opportunity to thank Patrick Burns and David Ratzan, who were at the conference but not presenters, for agreeing to add their chapters. I also encourage readers to consult both Shawn Graham's and Helen Cullyer's contributions as both are thought provoking in their own right. The former was written as part of reviewing the volume as a whole, for which I'm also grateful; the latter follows from both the SCS having been a partner in the event and from Dr. Cullyer having calmly facilitated the end-of-day discussion that included substantial input from the audience.

The idea of "discussion" is important to this volume. I believe I properly represent the intent of the contributors when I say that none of what follows is meant as a definitive and final statement. All of us recognize that digital tools introduce new tensions into the classroom and that those tensions often reflect, and are a subset of, tensions that exist more widely in society. Similarly, no single short volume and no single one-day conference can adequately range across all possible topics that can fall under the rubric implied by the title here. Using digital tools within the ancient world classroom is properly now under active consideration. As this volume was being prepared, *Teaching Classics with Technology* edited by Bartolo Natoli and Steven Hunt became available and readers will find much overlap between the chapters there and what follows here. A theme of the current volume is a sense that experimentation is good and that it comes with a risk of failure. Shawn Graham's new collection, *Failing Gloriously and Other Essays,* also published by the UND Digital Press, explores that topic in greater depth.

As we worked to incorporate the diversity that could fit within a single day, my co-organizers and I were able to include speakers from a range of types of institutions. Public, private, large, small, liberal arts and research oriented colleges and universities were represented. And there was gender balance, even if there may not have been full gender diversity as the term is now understood. Nonetheless, there is more work to do on that front so I am taking the opportunity of writing this preface to brainstorm a follow-up event that would build on the success of the first. Starting with possible titles, "Towards Digital Foundations for Diverse Ancient World Teaching" or "Building Diverse Teaching on Digital Tools for the Ancient World" are options that indicate the central place that diversity would have at the event. I would like the speakers to be diverse, again within the constraints implied by a one-day event. I would also very much like to de-center the civilizations of Greece and Rome. Coptic and Syriac (or Aramaic) should be present in the room not as adjacent subject to Classical Studies but as starting points in their own right. How might the efforts that these scholarly communities have undertaken to digitize themselves allow students to explore the Mediterranean and neighboring regions as places where the ability to navigate multiple cultural traditions was common and was an advantage? And as we ask students to explore these issues, what should we ask them to produce? I am comfortable that writing will remain an important mode of human communication, and I am equally comfortable with the idea that there are students who can express themselves visually with greater skill and comfort. But how do we grade a class where different students are using very different digital tools to produce very different end products? I presumed a decision that such a class would be a good thing, but I actually do not know that to be the case. I do presume that my students are neurologically diverse and I presume that there either is, or should be, similar diversity among those of us teaching the ancient world. How do we embrace that? Is there a useful overlap between the diversity of the Ancient World and the possibilities for diverse modes of teaching that digital technologies introduce? These are some of the issues that could arise and be discussed at a second DATAM-themed event. Again, it will take work and thinking, as well as listening, to turn these first ideas into a conference. But do stay tuned as I hope it can happen.

Works Cited

Natoli, Bartolo and Steven Hunt
    2019 *Teaching Classics with Technology.* New York, Bloomsbury Academic.

Graham, Shawn
    2019. *Failing Gloriously and Other Essays.* Grand Forks, ND, The Digital Press at the University of North Dakota.

Preface

Helen Cullyer, SCS

Digital Approaches to Teaching the Ancient Mediterranean was held just a few weeks before the 150th anniversary of a meeting, also held at New York University, at which the American Philological Association (APA), renamed as the Society for Classical Studies (SCS) in 2014, was founded. Although the founders knew nothing of "the digital", teaching was certainly at the forefront of their discussions. As the APA gradually transformed from an organization devoted to philology generally speaking to an association focusing on Greek, Roman, and ancient Mediterranean language, literature, history, and culture, digital approaches emerged as an area of concern in the mid- to late 20th century. There were task forces and occasional publications on word processing and the effects of computing on research and teaching. The association developed advisory boards for the new *Theasaurus Linguae Greaecae* and the venerable *L'Année Philologique*, as it transitioned from hefty print volumes only to an electronic database. The leadership of the APA was keenly aware that as research and teaching became more heavily reliant on digital resources, collaboration and coordination with academic institutions, libraries, and publishers both nationally and internationally was necessary.

More recently SCS has focused on open access resources such as the Digital Latin Library, hosted by the University of Oklahoma. It has also established a new endowment for the purpose of supporting the open access resources on documentary papyri accessible via papyri.info. SCS reaches the public via the open access SCS blog, edited by Sarah E. Bond and the Communications Committee, and also publishes digital project reviews on its website, an initiative started by Chris Francese. Finally, the annual meeting has become a venue for digital approaches to be discussed in a variety of different ways. Attendees are far more likely now than they were even a few years ago to find papers that utilize digital methods in regular paper sessions, while the all-day workshop Ancient MakerSpaces has become an important venue for sharing and demonstrating tools and resources and for building community. It is highly appropriate, therefore, that the SCS was able to co-sponsor Digital Approaches to Teaching the Ancient Mediterranean at the start of its 150th year.

On a personal note, I was delighted to be invited by the organizers to moderate the final panel discussion that drew together the threads of a wide-ranging day of talks. Prior to my appointment as the Executive Director of SCS, I was a program officer at The Andrew W. Mellon Foundation in the Scholarly Communications program and in that capacity reviewed and shepherded many grant proposals for digital projects and infrastructure. Much of our work as grant-makers involved working with libraries and publishers, but, as a former faculty member, I was always particularly interested in those projects that brought digital approaches to the classroom in order to engage students in a collaborative process of discovery, analysis, interpretation, and creation.

I hope that this volume will be attractive and useful to all those who are interested in teaching aspects, from material culture to language and literature, of the ancient Mediterranean, including those who identify as digital humanists and those who do not, and who, in fact, may be skeptical of what the digital can do for them and their students. If you fall into the latter camp, please read the following words of encouragement. The contributors to this volume take a capacious view of "digital approaches". The essays include discussion of important questions about equity in the classroom and the effects of the "digital divide" (Caraher) and about avoiding a naïve reliance on quantification and false objectivity (Walsh). David Ratzan engages with these difficult questions in his essay on information literacy, which is an increasingly important part of teaching in all disciplines. Further, you do not, and your students do not, have to be coding experts to engage in computational work, as Beaulieu and Bucci show. Finally digital approaches can be incredibly creative, as Blakeley shows in her work on gaming and Heath shows in his discussion of reconstructing ancient buildings. Despite the multiform use of digital approaches in the classroom that are discussed in this volume, Eric

Poehler argues "the digital objects that describe the ancient Mediterranean world are produced commonly in the context of research, but not commonly enough reoriented toward teaching." Let us hope that the use of digital objects, tools, and methods, as well as the engagement of students in dialogue *about* the digital, become more common in teaching the ancient Mediterranean. For the digital is here to stay.

Foreword

Shawn Graham

Once, says Roller, there was an onion. It was a *perpetual* onion.[1]

Once, in a faculty meeting, we debated what constituted "digital" work. "Why *can't* we just give a list of requirements, a list of things, and say, *this* is what digital work is?" asked my colleague, in some exasperation.

Once, in an interview, Joseph Weizenbaum (an important figure in the development of computing, as well as its deployment in the banking industry) said, the computer is the most conservative technology ever created. Instead of developing imaginative, *social* responses to problems, banks (for instance) could just throw computers at the problem. In this way, they did not have to develop regional or individual-level social responses to the problems they faced, they did not have to invent a *social* solution, and could instead continue to concentrate power and money at the center.

Once, there was an onion. An onion can be made to last forever, if you do it right. You only eat half of the onion. You put the bit with the roots on it in a jar of water. When the roots grow back, replant. Repeat.

The onion represents a breaking of how we're supposed to consume something. We don't have to consume *everything*. Instead, if we slow down a bit, we discover and can deploy a hidden affordance of latent life and make a renewable resource. Roller tells this parable at the conclusion of a piece examining the complicated material culture of 'machinic culture' of early 20th century Pennsylvania coal country.[2] The knowledge of how to make the perpetual onion was given to him by one of his ethnographic informants, a man who, with his lathe, was able to take something, anything, and make it into something else. He continually rejigged the available material into new

---

[1] Roller 2019
[2] Roller 2019

configurations, working against the ways consumer culture tried to order things. I grew up around people like that. My own grandfather for instance was constantly collecting things, tinkering with things, making old things do new work, making new things from old things. Growing up on a farm, you didn't buy *new* things. I remember my father welding bits of an old hay wagon to bits of an old truck box to make a new thing to do I do not remember what, now. More urbane folks might recognize something of the same creativity and failure-to-conform-with-consumerist-planned-obsolescence in the floruit of steampunk, and its political off-shoot, the right-to-repair movement.[3]

Teaching with/through/because of/despite "the digital" is the subject of this volume. The vision of digital classics in this volume is rather like that perpetual onion. And like my farming family, these scholar-practitioners take apart what the digital world offers us, recombining and reconfiguring the pieces in ways that are productive and perpetual. They are ways that crucially cannot be codified in a list of "what makes this work digital," as my colleague desired. And unlike the use of computers in education that Weizenbaum bemoaned as utterly conservative, that merely prolong the lifespan of perhaps outmoded ideas, not addressing the root issues and so lacking in imagination, the approaches described here *do* show those imaginative leaps that might enable us to teach the "partial puzzle analytics" that Walsh describes so eloquently. They are progressive in that they firmly put the students in charge. The current volume should be read in tandem with the 2016 volume *Digital Classics Outside the Echo Chamber: Teaching Knowledge Exchange and Public Engagement* edited by Gabriel Boddard and Matteo Romanello. The present volume feels more along the archaeological end of the Classics spectrum, while the earlier work is more along the philological end. Reading them together underlines the broader point made repeatedly in this volume that there are many ways to address the 'information gaps' about the Greco-Roman past, a multiplicity of voices and approaches: the digital contains multitudes.

---

[3] Lee 2019

Having enjoined you to go read the other volume too, I am going to cheat now and limit my comments to just what I see happening in *this* volume. This volume is a perpetual onion. Its layers fit together in a way that is extraordinarily useful for someone just beginning to get started in that part of the partial puzzle that digital approaches could help solve; one could consume part of the ideas in this volume, and find new ideas growing back to enrich us further.

That's probably enough of the onion metaphor, for now.

This volume provides examples, inspiration, caveats and warning—*hic sunt dracones*! Walsh reminds us of the value of Classics more broadly in an undergraduate education, and reminds us that digital approaches are just one of many approaches we can use; the warnings of "employability" and "truthification" should be well heeded. Folks who are digitally-inclined can fall prey to the same kind of chauvinism that exists for other tool-first mentalities, and Walsh gently reminds us to have "[...] a sense of respect and appreciation for individuals who had mastered different tools, or who were drawn to other puzzle pieces." Ratzan's contribution dovetails well with Walsh's, in that he demonstrates through some of his own digitally-inflected assignments how the "information semi-literacy" of students can intersect with the seeming objectivity of digital representations, and how our teaching can challenge that. His assignment on the seeming completeness of something like ORBIS—which students view and understand (at least at first) as having the same power and completeness of Google Maps—filtered through the lens of information literacy is particularly powerful.

Caraher's intervention is an important one for he reminds us (perhaps even alerts us for the first time) to the idea of first and second-level digital divides. Getting online is the source of the first-level divide. The students that he works with largely do not have access to the kind of (actual, literal, digital) bandwidth necessary to be "prosumers" or individuals who both create and consume digital culture (the second level divide). The original web browser contained within it the functionality to write to the web in the first place, but the commercial web split that ability out from the act of accessing information on the web. The "web 2.0" was all about having the tools that re-enabled people to create on and for the web, but that is also where the "second-level" divide comes from: these tools and software required a kind of access that cannot be assumed. The first level divide is merely getting online; the second level divide connects with Ratzan's concerns about the informational semi-literacy. Teaching in such an environment, and overcoming that second level digital divide moves us away from regarding "data" as something to be "mined" or extracted, but something that is co-created, something that is performed, in the course of asking questions. He also sees in this a way of being in the consumerist world that to my mind resonates with Roller's description of the archaeology of responses to "machinic" culture evidenced by folks in Pennsylvania's coal country. We build out of the pieces that the larger culture deigns to give us, repurposing those pieces to better (by our lights) ends.

The upshot of this kind of work is that it spills outside of the formal bounds of academe. We have a duty, do we not, to speak first to the people *outside* the discipline? That might be a contentious statement, but it also ties back to Walsh and Ratzan's concerns about the point of Classics in the 21st century. We have to be speaking, teaching, in the open (with the caveat that this can only be in terms of what it is safe for a person to do, given their local context). Burns recounts how the way early digital projects like Perseus and other projects spilled out onto the web in the 1990s enabled his engagement with Classics, and his ability to teach himself. "Digital Classicists have students who we never see, but whose studies are enriched by our work. Reciprocally, our field is enriched by their interest and participation, and this is a phenomenon worth noting" he says. In this, Burns is gently prodding us towards the same kind of ethic of generosity that animates Kathleen Fitzpatrick's 2019 *Generous Thinking*. The currents that animate digital approaches to teaching the ancient Mediterranean are wider and deeper than we might first have noticed! Blakely's work in archaeogaming continues in this vein. Video games with classical content or classical settings are extremely popular, and using such games in our teaching is becoming increasingly respectable, even commonplace.

What I find fascinating about Blakely's contribution is that it also demonstrates how the building of such a game, how the game iterates between building and testing/playing with students, is an act of research and co-creation of knowledge (in the ways that for instance Caraher described).

Videogames can be built without the act of coding; there are any number of platforms that allow one to build a game using graphical representations of the underpinning computational ideas (as in the Scratch language, for instance). No lines of code are necessary. Beaulieu and Bucci approach the problem of coding for Classics from a similar perspective, using a graphical user interface to plug together data workflows into which they pour data from various databases of Classical information. The point of teaching the students these workflows in this platform is to illustrate the ways our questions about the past can be represented in computational thinking. It is another instance of the same broad approach that Ratzan used to poke and prod at the ORBIS model. Once the habits of thought are instilled, the precise tool, the exact model, does not matter so much as the way of thinking to approach it. Partial puzzle analytics indeed! Heath and Poehler provide two more examples of what happens when we provide the "opportunity to allow digital tools to be part of the process by which students learn and think about the topic." In Heath's case, it is 3d models via photogrammetry, and how students learn to think about the materiality of the object and its situation in a nexus of relationships of other kinds of information. In Poehler's case, he uses the digital object, "... a product of the dual engagement of objects of inquiry and digital technologies [which] can face toward research or teaching equally and without contradiction."

These last three papers take us full circle to the question that Walsh opened with: what is the value added by digital approaches to teaching the ancient Mediterranean? The cycle of research and teaching described by the papers in this volume are anything but conservative, in the sense that Weizenbaum worried about. Digital work, digital technologies, are always changing; it's a moving target to try to pin down any one tool, technique, or approach as *the* thing we ought to be teaching with. The approaches in this volume are generous of spirit, productive and perpetual, and offer us actionable ways to integrate some digital inflection into our teaching, and for those of us already so inclined, ways to make the digital work we do more meaningful within our teaching.

Works Cited

ben-Aaron, Diana
    1985 Weizenbaum examines computers and society *The Tech* 105:16 https://web.archive.org/web/20191006185251/http://tech.mit.edu/V105/N16/weisen.16n.html

Boddard, Gabriel, and Matteo Romanello
    2016 *Digital Classics Outside the Echo Chamber: Teaching Knowledge Exchange and Public Engagement.* Ubiquity Press, London. https://www.ubiquitypress.com/site/books/10.5334/bat/read/

Fitzpatrick, Kathleen
    2019 *Generous Thinking: A Radical Approach to Saving the University* Johns Hopkins University Press, Baltimore.

Lee, John
    2019 What Ever Happened to Steampunk? *Modus* https://modus.medium.com/what-ever-happened-to-steampunk-4ac936905165

Roller, Michael
    2019 The Archaeology of Machinic Consumerism: The Logistics of the Factory Floor in Everyday Life. *Historical Archeology* 53:3-24 https://doi.org/10.1007/s41636-018-0149-0

Futures of Classics: Obsolescence and Digital Pedagogy

Lisl Walsh

This essay approaches the topic of digital pedagogies in Classics from the holistic perspective of the undergraduate Classics curriculum: what roles can digital pedagogies play, what is the 'value added' for students and teachers in terms of curriculum-wide learning goals, and, more broadly, how does the inclusion of digital-skills development within undergraduate Classics programs affect the preparation, diversity, and direction of the next generation of Classics scholars?[1] My perspective is informed, first, by my own experience in teaching basic statistics to undergraduates in the context of an intermediate/advanced Latin literature course, and, secondly, by the work accomplished by myself and other small-liberal-arts-college faculty for the 'Classics Tuning Project', one goal of which is to articulate the skills, methodologies, content areas, and habits of thought that we think form the essence and distinctiveness of an undergraduate education in Classics.

I want to begin by addressing the potential disappearance of Classics from the undergraduate educational landscape: in times of budget cuts and anti-intellectual cultural environments, what is it about this discipline—and specifically as it can be taught at the undergraduate level—that makes it worth keeping around? As part of the Classics Tuning Project,[2] I met with other small liberal-arts faculty

---

[1] I thank the organizers of this most excellent summit, David Ratzan in particular for devoting so much time to converse with me in the summer of 2018, and Sebastian Heath and Andrea Chang for their tireless attention to theme, representation, and logistics. Thanks also to Patrick Burns and Sarah Bond for their continued support, helpful critique and encouragement towards this contribution.

[2] For more information, there is a forthcoming article in *Classical Journal*; one can also find a brief summary of the project and its preliminary results (including our four Core Competencies), titled "'Tuning the Classics': Understanding the Under-

to try to define what it is that we are teaching when we say we teach "Classical antiquity." In the process of "tuning" (or tuning into) our learning outcomes, we also wanted to answer the above question and provide language for those looking to defend the discipline's worth.

In the following, I would like to present my personal hypothesis—stemming from one of the Tuning Project's four Core Competencies, "Interdisciplinarity," and from the results of the Project's alumni survey—as to how the discipline of Classics can both distinguish itself from other undergraduate programs and market itself as a course of study that prepares students exceptionally well for their occupational futures.

Unlike undergraduate curricula in English literature, History, or Modern/Foreign Language, a Classics curriculum often necessitates an acknowledgement of the relative scarcity of its evidence and the consequent necessity for multi-disciplinary approaches to the evidence we do have. The skill this trains, I propose, could be called "partial-puzzle analytics." The term refers to an approachable analogy I like to use to explain how studying Classics simultaneously trains students in skills of precision and/in analysis *and* outside-the-box creativity for/in problem-solving: that studying Mediterranean antiquity is like trying to put together a 500-piece puzzle when we only have five pieces and no picture to guide us.

First, the micro-analysis: what we observe about each of those puzzle pieces has ramifications both for how we will view the other pieces and for the resulting image of the whole puzzle. (Are we going to pay attention to the color yellow in the pieces? What picture does that imply? Or are we going to notice the sharpness and fuzziness

---

graduate Curriculum," in the July 2018 *SCS Newsletter*: https://classicalstudies.org/publications-and-research/scs-newsletter-july-2018-tuning. The full list of Tuning contributors, our meeting agenda, and the scope of our grant from the Associated Colleges of the Midwest is available at https://www.acm.edu/professional_development/project/62/tuning-the-classics. I owe a debt of gratitude to Clara Shaw Hardy, John Gruber-Miller, Angela Ziskowski, and Sanjaya Thakur for letting me join and contribute to this project, to the Associated Colleges of the Midwest (especially Brian Williams) for their generous financial support, and to the many participants at the 2019 Annual Meeting of the SCS and the 2019 ACL Centennial, whose comments and questions have informed my thoughts herein.

on each piece, and try to construct a picture that way? Do those two approaches help narrow our reconstruction options, or do they create mutually exclusive images?) In teaching the ancient Mediterranean, we teach students how to beat each piece of evidence with any and all manner of approaches such that it will yield as much information as possible.[3]

This necessity of micro-analysis of puzzle pieces through any and all means is one of the ways of thinking that is, if not unique to Classics, then certainly a hallmark of an undergraduate education in it. The metaphor is also a good way to communicate to non-Classics audiences what we mean when we say that Classics must be "interdisciplinary:" the successful (re)construction of the whole puzzle requires us not just to see every minute detail of each piece (what some might call "critical reading"),[4] but also to see simultaneously how the interpretive frames we use as well as the information they engender *also* affect (and are affected by) the other pieces and the larger puzzle picture. In more succinct language, the Core Competency of "Interdisciplinarity" from the Tuning Project expects that successful undergraduate Classics students should be able to ask questions about evidence from more than one disciplinary perspective, to understand the advantages and limitations of more than one disciplinary approach, and to synthesize holistic arguments about sources by making use of more than one disciplinary perspective.

Clearly, "digital" approaches, in the context of interdisciplinarity, provide useful ways of looking at the evidence, but are they *more* important or *more* useful than any other approach? Yes and no. Much like the disciplinary lenses of philology, or gender studies, or sociology, or archaeology, or economics, or geography—each of which

---

[3] Another metaphor here is the piece of evidence as a piñata full of candy, where different methods of analysis result in different "candy" coming out of the evidence. I use the word "beat" intentionally, because analysis is a violence done to the source, and we should own that it is an imperialist attitude towards the sources that make us treat them as though they ought to be open to us.

[4] Here I am thinking of George Anders' *You Can Do Anything: The Surprising Power of a "Useless" Liberal Arts Education* (Little, Brown and Co., New York, 2017), who points out the transferrable value (i.e., in terms of employability) in having been trained to read 'between the lines' of a text (pp. 35-45, 106-28).

can generate different questions for the sources that yield different results, and each of which has its own "blind spots"—so too any specific digital approach to the sources will generate different questions, different data, different "blind spots," and different conclusions. In this respect, they are intensely useful for the scholarly community (including undergraduates) to yield new information from our sources and thus be differently able to (re)construct the "big picture" of the ancient Mediterranean.

But if the undergraduate curricular goal is that students will be able to "do partial-puzzle analytics"—i.e., to "think like a Classicist"— then digital approaches are *applicable* but *not essential*. In learning the mental habit to perceive and retain multiple, differently valid permutations of the realities of antiquity based on the evidence we have, the tools through which students learn and practice this habit are not what's important.[5] Rather, it is being able to work with sources *using multiple tools*, and, indeed, to see how a multi-tool approach makes the puzzle pieces evade a singular appearance; students should reckon with the logical contradictions and even paradoxes that arise from the different knowledges generated in the process of multi-tool analysis. Hence, to the extent that any digital approach (whether that be mapping, image-manipulation, tree-banking, or, as I taught in a "Seneca and Statistics" course,[6] simply counting up words) will give students another means for source-analysis and foster the challenge of synthesizing results of each perspective into something meaningful about the source itself, digital approaches should by all means be incorporated into an undergraduate curriculum. But they do not,

---

[5] This is of course a controversial statement, especially when one sees, as the Tuning Project does, ancient languages as one of many "tools" that can be used to teach "Interdisciplinarity." It should be said that "digital" approaches as such are also not said to be essential in the Tuning Project Core Competencies. One could argue, however, that the mental act of creating and suspending multiple permutations of possibility is a skill that is honed rather efficiently in the process of learning how to translate ancient languages.

[6] It would unacceptably lengthen this essay to go into detail on this course; for the curious reader, one can find the problem sets and syllabus at https://beloit.academia.edu/LislWalsh/Teaching-Documents.

and should not, I think, intrinsically supervene or replace any other disciplinary approach already present in the undergraduate Classics landscape.

The other, equally important half of "partial-puzzle analytics" is thinking about the evidence in relation to the "big picture." Unlike many other historical time periods, other non-Anglo-American cultural studies, or other bodies of literature, however, the "big picture" of the ancient Mediterranean has a lot of empty space. Thus, the mental process of seeing the interrelationship between the one piece of evidence and a larger environment that is full of unknowns is all the more challenging. With few anchors holding the puzzle pieces in place or determining what does and does not appear in the "big picture," students practice, again, seeing multiple possible permutations simultaneously, seeing the dynamic and inherent interdependence between the evidence and its environment, and they practice the arduous task of "bopping" easily and quickly between the micro-analysis of the evidence and the big picture. This practice engenders two of what (I think) are the other strengths of the undergraduate Classics curriculum: (1) being able to perceive and infer patterns from what seem to be (for many other disciplines) threadbare data sets, and (2) being able to "bop" from tiny details to big pictures.

With these habits of mind built from an interdisciplinary approach to scarce resources, a student who can do "partial-puzzle analytics" can to it with any puzzle and any tools. Again, learning "digital" tools can work, but they are not essential. A Classics student trained to use philology and gender studies, I claim, will just as easily be able to transfer this skill to another setting: taking a big picture to make an inference about a tiny thing, taking a tiny thing and pulling out tons of information from it, seeing the patterns in data sets and finding connections between seemingly disparate pieces of evidence. This student will have instincts to believe that a single approach to a data set is not sufficient, to learn new tools with which to examine evidence, to learn the limitations and knowledge gaps produced by each tool, and to be creative about what other tools they might be able to apply (or other questions to ask) in order to generate a nuanced synthesis that is as comprehensive as possible.[7]

Finally, and not least importantly, Classics distinguishes itself in that students can learn how to use the evidence to reckon with and (re)construct the metaphorical 495 missing puzzle pieces.[8] I think we as Classicists underestimate the intellectual rigor that can develop from acknowledging the evidentiary gaps and teaching students how they themselves must (in order to "think like a Classicist") learn how *creativity and imagination* play essential roles in any attempts made by the scholarship to explain anything about the ancient Mediterranean. In doing so, we are teaching not just "lateral thinking" (Anders, *op. cit.*, p. 38, cf. "nonlinear thinking," p. 68), or 'outside-the-box' problem-solving, but also "world-building"—a skill that is increasingly necessary, I think, for understanding the contingent nature of present reality and for creating change and innovation for the future.

So What's Wrong with "Digital"?

This to me is part of the obsolescence of Classics (and by obsolescence, I mean more the etymological "getting in the way," or "being an obstacle"): as a mode of thinking about the use of the discipline in undergraduate education, the Tuning Project's understanding of Interdisciplinarity rejects the idea that any one tool is indispensable for an individual to "do Classics." In an increasingly digital society where, clearly, some tools seem more worth learning than others, this

---

[7] I refer readers to Mona Hanna-Atisha's *What the Eyes Don't See: A Story of Crisis, Resistance, and Hope in an American City* (One World, New York, 2018), in which the author credits not only her own background in humanities but also her habit of reading widely for helping her realize that mapping the lead levels of tested children in Flint, MI would supply the numerical legitimacy needed to support a claim that there was excessive lead in their water. Here, the "tool" of mapping the data needed to be used alongside the "tool" of binomial statistics to reveal results that were otherwise hidden by the latter.

[8] And who says there are (metaphorically) 495? Part of the challenge is that we can never know exactly how much we don't know. I use the term "(re)construct" purposely: see discussion of "Digital Pitfalls" below.

assertion that someone who approaches Vergil using gender studies and philology is doing the same thing as someone who analyzes Minoan-age Crete using art history and economics or the same thing as someone who analyzes the Roman army using an English translation of Tacitus' *Germania* and numismatics flies in the face of both the history of the discipline itself and its perceived utility outside of academia. Classics as a discipline is still, I think, living with the ghosts of Bentley and Mommsen: this false idea that the ancient Mediterranean is a closed body of knowledge, with a few necessary tools that any professional in the field ought to have up in their brains at all times. Similarly, we act as though our undergraduate and graduate programs will provide a comprehensive training in all the tools of the discipline: ancient languages, numismatics, historiography, philology. But we all know that every program has different strengths, can teach certain tools well and not others. What the Tuning Project's perspective tells *me*, at any rate, is that this is just fine, and that it is in fact the valuing of certain tools over others that will help the discipline become obsolete, because doing Classics is learning how to do "partial-puzzle analytics," and it involves the mastery of more than one tool, but not the mastery of particular tools as their own ends.

So, what's wrong with using digital approaches, or at least teaching our students how to do Classics with a tool that seems explicitly transferrable, pragmatic, and relevant to the present? Nothing! I do, however, think that we need to flesh out some potential pitfalls if and when we as teachers contemplate making a "digital turn" in our teaching. Both of the following, I think, have been proposed as ways of fighting *against* the potential fading away of Classics in higher education, but I want to argue that, if improperly deployed, they actually help make our discipline a redundancy in the undergraduate landscape.

Pitfall 1: Classics and/under "Employableism"

One reason we might teach tools that are right now perceived as "useful" is because we worry about students finding employment with an undergraduate degree in Classics. We want to help them (and ourselves) by teaching and marketing Classics as a means to learn tools that will secure employment. We think, if we teach them Python or ArcGIS, they will at least have a skill to market. This move seems like it is making the discipline more relevant, more desirable, because administrators still need places to send students who need to improve their writing and reading, parents love the promise that a STEM-ified College degree seems to make, of instant jobs that pay well, with upward mobility. But what then separates Classics from a rhetoric and composition course, or a programming course? In a utilitarian and zero-sum market of streamlining budgets, why keep Classics around if students can learn the tools it teaches elsewhere? And for those programs who are still teaching Classics with relatively "useless" tools, it seems all the more difficult to justify them as worthwhile expenses.[9]

One way to subvert this disaster is to focus on "partial-puzzle analytics" as the endgame of an education in Classics. We must remember that using the ancient Mediterranean as a means to learn a tool is not the same as learning the tool in order to practice "thinking like a Classicist," and we must remember that no tools are inherently more useful than others for teaching this habit of mind in the (relatively) threadbare archive of Classical antiquity. We should consider the time students and instructors will invest in learning a given tool, how that tool will be worked into the curriculum as part of students' sustained multi-tool analysis of evidence, and how that tool (or the time invested) affects the students' abilities to practice working with the *gaps* in the evidence as well as with the evidence itself. So I say, if you as an instructor and researcher *already use* Python or ArcGIS, if you are used to thinking with these tools (their limitations, biases,

---

[9] And who is to say that the digital tools we are teaching now will be useful in 20 years, either for the students or the teachers who have mastered them? Knowing how to code, as we learned in 2001 and again in 2008, is not a "recession-proof" skill.

etc.) as part of your own multi-tool analysis of ancient evidence, then by all means have students practice using this tool as part of their curriculum.

Pitfall 2: Truthification and "Big" Data

The gist of this pitfall is that, as we get larger and larger data sets of texts that are now digitized, as digital tools give us insight into aspects of our evidence heretofore inaccessible, and as we use tools to gather even more and more precise information about material remains, it becomes easier to focus exclusively on the evidence itself and to forget its context. When data sets are large, working on them takes up relatively more attention and focus, and it is easy, again, to lose sight of the relationship between the (still relatively small) data set and the (still relatively empty) big picture—one of the essential elements that, I have argued, distinguishes a Classics education. When teaching with online databases like Phi or Perseus (which I use in my own research and therefore feel comfortable incorporating into class assignments),[10] it is crucial to counterbalance the analysis of evidence with exercises that force students to think about contextualization, the "big picture," and the creativity involved perceiving and filling in the many gaps in the archive.

Relatedly, digital tools for analyzing data give off the veneer of rigor, science, and objectivity. It becomes easier to believe that the data set is comprehensive, and harder to see its limitations and biases (to say nothing of the limits and biases of the tools themselves). Why is this a problem for the discipline? For me, it comes back to knowing how to use your tools properly, and knowing the limits and biases of the data being generated. One forgets that data, numbers, are generated by humans. Even in the natural sciences, there is no such thing as an objective observation, an objective gathering of data. A person, with all their assumptions, is always behind the questions getting asked, the measurements being taken, the numbers being included

---

[10] https://latin.packhum.org/; http://www.perseus.tufts.edu/hopper/help/quickstart.jsp.

or tossed out, the tests to which the data are subjected, and the conclusions that are drawn.[11] There is no such thing as "raw numbers" or "raw data." In using and teaching these tools, in this age of trusting in the Truths to be gained from more data, it is easy not to teach "partial puzzle analytics"—the realization that all this data is a tiny puzzle piece and is interdependent on the assumptions made about what the rest of the puzzle looks like.

Finally, digital tools can make it more challenging to realize that any solution produced by the data and the tool is—because of the still-missing gaps in the archive—merely one of *several* possible and equally valid (re)constructions of the evidence and its relation to the big picture. It is so easy to forget how much conjecture, how many assumptions, go into the creation of such things, and how *relatively little* of the ancient world is actually "Known." Digital tools fool us—and our students—into thinking that there could exist one correct version of the puzzle and that their job, our job, is to "find it." But there is no finding it because even if we had all the evidence of antiquity we wanted, on some level we know that evidence cannot stand in for reality.[12] So, against the seeming increase of certitude that stems from digital tools, and against a teleological narrative of this discipline coming ever closer to "true knowledge," I want to posit that the study of the ancient Mediterranean is useful and should continue precisely because there will always be multiple "correct" versions of the puzzle. Seeing the "actual" version is not the point; to the contrary,

---

[11] For this, I refer readers to Duana Fullwiley's 2008 article, "The Biologistical Construction of Race: Admixture Technology and the New Genetic Medicine" (Social Studies of Science 38(5): 695–735).

[12] To what ends is the discipline of Classics, and to what ends are digital tools pushing the discipline? I see them framing the discipline as a pursuit of What Actually Happened—as if Classics as a discipline could stop once everyone agreed what the puzzle looks like and how all the pieces fit in. I see this conundrum as part of an "original sin" of Classics as a discipline, built on the idea that logic and science would lead us from our archive to the omega-text. In another piece, I discuss the disservices this driving force has done to the research on Roman women and Senecan tragedy.

what serves our students best is the mental challenge of trying to understand relationships between scant evidence and the reality we and they construct around it. *This* is how they learn to be world-makers.

Final Thought: Future of the Discipline and What needs to Change

In October 2018, I informally polled a bunch of fellows at the American Academy in Rome, asking them where they thought the future of the discipline was going and what they thought it would look like in 50 years. I was thinking specifically of how the discipline was taught, but we talked mostly about research. One said "Classics is going to move later, and it's going to move East... everyone's going to have to learn Coptic," and "I like Vergil as much as the next guy but, we don't need another dissertation on Vergil." A visiting scholar said papyrology will be more important than it is currently, that digitization and access will be increased (but will not necessarily yield more answers without a human to interpret the data), and that the field would not die out, at least not in Europe, because as she said "it is our cultural heritage." Two archaeologists both said that data is going to be bigger and will provide so much more information than we used to have—about ancient people, ancient sites, etc., but there was concern that archaeologists will have to turn "back" toward writing narrative, explanatory pieces that actually build stories for the data. As one explained, data published without the interpretation *of the excavator* is not going to be as helpful as we think it will be. (She also pointed out that a lot of sites, sources, will in 50 years be underwater, which made everyone at the table laugh nervously.)

I continue to be struck by what seems to me to be an allegiance to their tools and/or their puzzle pieces as being more useful than others. What is missing, it seems, is a sense of respect and appreciation for the individuals who had mastered different tools, or who were drawn to other puzzle pieces. Why bother trying to explain Coptic to a Vergilian philologist if they were not going to learn it themselves? The *discipline* was going to shift to Coptic, or excavator narratives, or big data. Maybe readers of this volume have a sense that the *discipline* is going to shift digital, that "non-digital" Classics will die out. But it seems to me that, once we admit that individual Classicists and individual programs cannot master or teach every single tool in the arsenal, and once we also admit that we need *representative* masters of *each* of those tools in order to get the best views of those puzzle pieces and (if we want to get all scientific method about it) to eliminate puzzle reconstructions that are probably wrong, we should strive to foster and value programs and departments that are teaching "partial puzzle analytics," no matter what tools they are using. We need to respect people who use different tools than we do, and we should learn how to explain our tools to others in such a way that we can collaborate more effectively for research publications and investigations. Finally, we should continue to *hire* people for their tool mastery, and while we might aim for a breadth of tools to be represented in any given department, we should recognize that the tools in and of themselves are not what trains a Classicist to think like a Classicist.

Works Cited

Anders, George
   2017 *You Can Do Anything: The Surprising Power of a "Useless" Liberal Arts Education.* Little, Brown and Co., New York.

Fullwiley, Dana
   2008 "The Biostatistical Construction of Race: Admixture Technology and the New Genetic Medicine. *Social Studies of Science* 38(5): 695-735.

Hannah-Athisha, Mona
   2018 *What the Eyes Don't See: A Story of Crisis, Resistance, and Hope in an American City.* One World, New York.

Walsh, Lisl
   2018 'Tuning the Classics': Understanding the Undergraduate Curriculum. Electronic document, https://classicalstudies.org/publications-and-research/scs-newsletter-july-2018-tuning.

Teaching Information Literacy in the Digital Ancient Mediterranean Classroom

David M. Ratzan

I. Introduction

A familiar rite of passage for early-career academics and academic librarians with instructional responsibilities is the Statement of Teaching Philosophy. Are these compulsory études effective when it comes to landing a job as an assistant professor or academic librarian? *Quot docti, tot sententiae.* Can they be valuable in themselves as theoretical meditations on a practice that lies at the heart of the academic mission? Surely Plato thought so. It is unfortunate and unfortunately unsurprising that most of us who teach the ancient Mediterranean world to undergraduates are typically asked to reflect on our craft as teachers only once or twice in our careers: once at the very beginning, when applying for jobs; and then many fewer of us when submitting tenure or promotion dossiers. Yet it may be to this state of affairs that we owe the richness and energy which characterized the discussion at the conference that generated these proceedings, in which seasoned educators from very different kinds of departments and institutions across the United States engaged critically with their experiments in and experience of digital approaches to teaching the ancient Mediterranean world. During the presentations I found myself returning to and reflecting on my own Statements of Teaching Philosophy, written long before I had taught any of the workshops and lecture classes that now form the staples of my teaching, and how digital resources, models, and computational approaches have changed what and how I teach, and why.

Fresh out of graduate school I wrote that I attempted to plan classes with three nested pedagogical objectives: a *subject* lesson (e.g., what kind of text is the "Oracle of the Potter" and what does it actually say?);[1] an *object* lesson (e.g., what does it mean to read the "Oracle of the Potter" as "resistance" literature in the Ptolemaic or Roman Empires?); and what I called a *take-home* lesson (e.g., what are our own contemporary forms of "resistance?" How are they culturally and historically conditioned, and what does that mean for the discourse of and potential for "resistance?"). This is still the way I approach lesson planning; yet over the past five to six years I have increasingly found myself incorporating a new pedagogical objective into some classes, one intimately bound up with the project of teaching antiquity in our digital present and informed by the information literacy pedagogy of my library colleagues. A focus on information literacy may seem to intersect only obliquely with the theme of these proceedings. First, information literacy is, of course, a wider and more general competency, one we might hope that all undergraduates attain, not just those studying the ancient Mediterranean. Second, it is also not a domain restricted to specifically digital resources and approaches to information. I concede both propositions; yet I nevertheless hope to show in this contribution that designing activities and paper topics with information literacy in mind can help to lay a foundation for critical engagement with digital approaches as well as to adumbrate for a non-specialist, undergraduate audience the distinctive challenges, pleasures, and intellectual value of studying the ancient world. In the next part of this essay (Section II), I will review the recent (and to my mind salutary) shift in the theory and practice of information lit-

[1] The so-called "Oracle of the Potter" is an Egyptian apocalyptic-oracular text, most likely written in demotic in the third century BCE in reaction to Ptolemaic rule. Fragmentary versions of the text survive only in Greek in five papyri, all from the Roman period (late second-early third century CE, and so clearly continuously read and re-read in different political and social circumstances). Still fundamental are Koenen's basic studies (1968; 2002). An English translation (which does not reflect Koenen 2002) may be found in Kerkeslager 1998. Recent noteworthy studies on the text and the basic question of revolt in Greco-Roman Egypt include: Collins 1994; Potter 1994: 192-206; Beyerle 2016; Gruen 2016; Ladynin 2016; Ludlow and Manning 2016; and McGing 2016.

eracy in the United States. In the final part (Section III), I will describe specific projects I have assigned in class that include an information literacy objective.

II. Information Literacy and the ACRL Framework

There is a tremendous amount written about information literacy on both a theoretical and practical level.[2] Very little of this literature, however, is directly pertinent here, since most of it addresses the challenges of teaching information literacy per se (i.e., independent of any specific discipline) and the specific instructional role and responsibilities of libraries and librarians. What is worth noting here, particularly for teaching faculty, is that this field has witnessed a recent and noteworthy development, one which is still percolating through the academy. In 2015 the Association for College and Research Libraries (ACRL), a division of the American Library Association, published its new *Framework for Information Literacy for Higher Education* (the "Framework"). The Framework entirely replaced its predecessor, the *Information Literacy Competency Standards for Higher Education* (the "Standards"), which had been approved by the ACRL Board of Directors in 2000 and subsequently adopted by several other organizations and state legislatures and implemented widely as the basis of information literacy curricula and courses across the United States. The Framework is not an update and revision of the Standards but instead a complete reconsideration of the theoretical basis and pedagogical strategy of teaching information literacy.[3] Unless you are particularly interested in information literacy or closely connected to an academic library's instructional program, this was a revolution that very likely passed you by. The irony (and one not lost on many librarians) is that teaching faculty may be better placed to do some of the work of this revolution than librarians.[4]

---

[2] I would like to thank Lauren Kehoe, Michelle Demeter, and Jill Conte for generously sharing their perspectives and suggestions about teaching information literacy in the wake of the publication of the ACRL Framework.
[3] See, e.g., Oakleaf 2014 and Foasberg 2015.
[4] "Framework" 2015: 7, 27-28; cf. Wilkinson 2016d. Bombaro 2016 is highly

The Framework defines information literacy as:

> the set of integrated abilities encompassing the reflective discovery of information, the understanding of how information is produced and valued, and the use of information in creating new knowledge and participating ethically in communities of learning. (2015:8)

Likely, this will seem reasonable to pretty much anyone teaching in secondary or higher education today. However, as Marcus Leaning relates in his history of the concept, the content and the aims of information literacy have changed dramatically over the last three to four decades. Information literacy was one of several new "literacies" discovered and articulated in the second half of the twentieth century, with the first attestation of "information literacy" appearing in 1974.[5] From the start, information literacy has been connected conceptually to technological development, the growth in the amount and types of information available, and the multiplication of ways in which it is created, packaged, discovered, retrieved, delivered, and now increasingly shared and reused. Pedagogically, the focus has, until quite recently, been very much on the teaching of the technical skills associated with specific tools or resources. In some ways, the culmination of this phase was the erection of the Standards. This document identified five standards, 22 performance indicators, and 87 (!) outcomes for the information literate. To give an example:

> Standard 2: The information literate student accesses needed information effectively and efficiently.

---

critical, but behind the palpable anxiety lies a hard reality of the challenges that face librarians trying to work with teaching faculty to put the Framework into practice. Recent work on the impact and implementation of the Frames is generally more positive and optimistic about collaboration with teaching faculty: e.g., Dawes 2019; Dolinger 2019; Latham et al. 2019.

[5] Leaning 2017: 40.

Performance Indicator 2.3: The information literate student retrieves information online or in person using a variety of methods.

Outcome 2.3b: Uses various classification schemes and other systems (e.g., call number systems or indexes) to locate information resources within the library or to identify specific sites for physical exploration. (2000: 10).

If your institution teaches information literacy classes, it is a good bet that the curriculum was, and may still be, based on these standards.

The Framework is a different animal. Its prologue asserts that "the rapidly changing higher education environment, along with the dynamic and often uncertain information ecosystem in which all of us work and live, require new attention to be focused on foundational ideas about that ecosystem" (2015: 7). Accordingly, it dispenses altogether with the idea of standards defining some objective technical proficiency in favor of six "Frames":

- Authority Is Constructed and Contextual
- Information Creation as a Process
- Information Has Value
- Research as Inquiry
- Scholarship as Conversation
- Searching as Strategic Exploration

Before diving into what these Frames mean and how they can be helpful in teaching the ancient Mediterranean world, it is important to explain their intellectual foundations, specifically two educational theories: "threshold concepts" and "metaliteracy."

There is now an exhaustive monograph dedicated to metaliteracy, but for our purposes the basic idea suffices: it denotes the extension of traditional information literacy skills (e.g., determine, locate, access, understand, use, cite, etc.) to the more fluid, dynamic, and social information ecosystem we now inhabit, in which users collaborate, participate, produce, share, and reuse information.[6] Behind the jargon lies an important reality: these new modes of creating, assembling, consuming, and sharing information have important implications for data and interpretation; and our students need to learn not only to appreciate these implications, but also to adopt a more active, critical stance with regard to their intellectual and ethical participation in these living networks of information (which the theorists call "metacognition").[7] The Framework is an attempt to reorient the teaching of information literacy along these lines, to cultivate the skills and critical habits of mind required to navigate our world of interactive and recombinant information. I will return to some of these points below when I discuss working with papyrus documents from Ptolemaic and Roman Egypt.

"Threshold concepts" are a cottage industry unto themselves in educational theory and seem to have reached the pitch of their popularity in first half of this decade, just as the Framework was being drafted.[8] The basic premise is that each field of inquiry has a set of core concepts, which, once taught, are *transformative* (they precipitate a radical change in perspective), *irreversible* (they are hard to "unlearn"), *integrative* (they expose a deep interconnectedness of phenomena or thought patterns in a particular discipline or methodology), *bounded* (they are specific to a discourse or field or method, or perhaps better put, they are the foundational, constitutive ideas or paradigms that define a discourse or field or method), and potentially *troublesome* (they may be counter-intuitive, hard to internalize or operationalize, run counter to deeply held views about the world, etc.). To learn these ideas is in some sense to learn to "think like" a physician, an economist, a historian, an archaeologist, a classicist, etc.

---

[6] Mackey and Jacobson 2014; cf. Mackey and Jacobson 2011 and 2016.

[7] See Caraher's incisive contribution in this volume for some of the challenges, limits, and perhaps unwitting ways in which educators serve corporate interests when trying to teach their students to be "prosumers," or participants in the digital world who consume and produce "products," "content," and "media."

[8] The seminal article is Meyer and Land 2003, further elaborated in Meyer and Land 2005 and 2006. For the application threshold concepts to information literacy in libraries in advance of the drafting of the Framework, see, e.g., Townsend et al. 2011.

By way of example, consider the change in perspective and subjectivity that takes place when a student learns how a classical literary "text" is constructed, and therefore what classicists mean when they speak about the "text" of, say, Plato's *Republic*. There is, in a real sense, no going back once the veil has been lifted and she understands that every classical text is the complex and inherently unstable product of an evolutionary history of composition, publication (which was itself very different in the ancient world), copying, recopying, correction, collation, and, finally, modern scholarly intervention and printing (our books look very different from the manuscripts the ancients read). To see any particular text as but one possible instantiation of a tradition and a process, and so unlike almost all texts written in the last century, forces a change in perspective with respect to what that "text" is—and indeed what any and all classical literary "texts" are. It also establishes a different relationship between the reader and the text. True, the responsibilities and engagement now demanded of the initiated reader can be suppressed for casual reading, but they can only be pushed off: it would be virtually impossible to forget or unlearn this new understanding or not to engage with it when embarking on a "serious" reading of any classical text. Again, such knowledge is integrative, in that one now sees and can therefore abstract the processes involved in the creation and editing of all classical texts. Similarly, philology in all of its varied facets, from grammar to diction to stylistics to socio-linguistics, is revealed to have a motivated, dynamic, constructive—and therefore potentially circular—relationship to the texts and language it purports to describe. All of these revelations are potentially troubling: the text is no longer unitary and simple (if any text is); one can take neither the text nor "reading" for granted; the ap. crit. (and the abbreviation is a shibboleth of one's membership) acquires a meaning and a function—and if you know what it is for, you also know that it is not necessarily to be trusted, since it is the creation of an editor and thus rests on (usually still) his authority—which is based on what? Ultimately, the appreciation of his ability to think like a philologist and an editor in the eyes of other Classicists. Is Classics a "profession" or a "discipline"? Well, to approach texts in this way is absolutely the hallmark of a philologist of premodern texts, if not Classicists alone: it is one of the ways of thinking, perhaps a threshold concept, that defines or bounds what Classicists do *qua* Classicists—an idea to which I will return in the conclusion.[9]

Since the introduction of threshold concepts, there have been those who, perhaps predictably, have sought to identify and compile definitive lists for their respective disciplines, and the Framework is just such an attempt to define (at least some of) the threshold concepts for information literacy (which presupposes that information literacy is itself an independent discipline, asserted by the Framework but the subject of some debate).[10] As with many theories, this one has been applied mechanically, as some have debated how many of the italicized qualities above have to be valid, and to what extent, in order for a particular concept to qualify as a "threshold concept."[11] While such a discussion may have the salutary effect of pushing practitioners to clarify precisely what it is that they do when they do it (and here I cannot help but think of Stanley Fish's classic essay, "What makes an interpretation acceptable?"), erecting some disciplinary cannon of threshold concepts seems as unnecessary as it is quixotic, if only because methods and disciplines change over time, and individuals can and surely will find that different concepts spark some set of the important transformations contained in the ideal type of the threshold concept.[12] To my mind, one realizes the pedagogical value of threshold concepts by resisting the temptation to dogmatism and instead seeing the idea as a convenient label for a bundle of qualities that reflect a certain educational rite of passage that most teachers recognize and strive to catalyze (in fact, we might see Plato as the original threshold concept theorist). In my assignments, I therefore look to stimulate elements of the threshold experience, which I have

---

[9] See Walsh's contribution in this volume and her identification of "partial puzzle analytics" as something like a threshold concept in Classics. I engage with her critique in the conclusion to this piece.

[10] See, e.g., Wilkinson 2016d.

[11] E.g., Wilkinson 2014a.

[12] This is, to my mind, one of Fish's central messages, *avant la lettre liminaire*, when it comes to present and potential future rules of interpretation.

found pedagogically valuable, without concerning myself with the need to credential any particular idea as a "threshold concept" per se in either information literacy or any branch of ancient studies.

Below are the six Frames and their explanations, presented in their original, alphabetic order. For ease of reference, however, I am numbering them. The ideas embodied in the Frames will be familiar to almost anyone who teaches. In fact, that is the point: to distill and articulate what we do as twenty-first-century scholars when it comes to the critical discovery and use of relevant information in our research and writing, even if we as teaching faculty do not necessarily think of this as teaching "information literacy" when we model these practices and dispositions.[13] In the Framework, each Frame is followed by a list of associated *knowledge practices* (basically, skills) and *dispositions* (the new metaliterate subjectivity that attends the threshold experience), which I have omitted. If you read the Framework (and it is refreshingly succinct), I recommend reading Lane Wilkinson's trenchant criticism of just about every aspect.[14]

1. Authority Is Constructed and Contextual: Information resources reflect their creators' expertise and credibility, and are evaluated based on the information need and the context in which the information will be used. Authority is constructed in that various communities may recognize different types of authority. It is contextual in that the information need may help to determine the level of authority required.

   Experts understand that authority is a type of influence recognized or exerted within a community. Experts view authority with an attitude of informed skepticism and an openness to new perspectives, additional voices, and changes in schools of thought. Experts understand the need to determine the validity of the information created by different authorities and to acknowledge biases that privilege some sources of authority over others, especially in terms of others' worldviews, gender, sexual orientation,

---

[13] Cf. Dawes 2019 and Latham et al. 2019.
[14] Wilkinson 2014a-2014g, 2016a-2016f.

and cultural orientations. An understanding of this concept enables novice learners to critically examine all evidence—be it a short blog post or a peer-reviewed conference proceeding—and to ask relevant questions about origins, context, and suitability for the current information need. Thus, novice learners come to respect the expertise that authority represents while remaining skeptical of the systems that have elevated that authority and the information created by it. Experts know how to seek authoritative voices but also recognize that unlikely voices can be authoritative, depending on need. Novice learners may need to rely on basic indicators of authority, such as type of publication or author credentials, where experts recognize schools of thought or discipline-specific paradigms.

2. Information Creation as a Process: Information in any format is produced to convey a message and is shared via a selected delivery method. The iterative processes of researching, creating, revising, and disseminating information vary, and the resulting product reflects these differences.

The information creation process could result in a range of information formats and modes of delivery, so experts look beyond format when selecting resources to use. The unique capabilities and constraints of each creation process as well as the specific information need determine how the product is used. Experts recognize that information creations are valued differently in different contexts, such as academia or the workplace. Elements that affect or reflect on the creation, such as a pre- or post-publication editing or reviewing process, may be indicators of quality. The dynamic nature of information creation and dissemination requires ongoing attention to understand evolving creation processes. Recognizing the nature of information creation, experts look to the underlying processes of creation as well as the final product to critically evaluate the usefulness of the information. Novice learners begin to recognize the significance of the creation process, leading them to increasingly sophisticated choices when matching information products with their information needs.

3. <u>Information Has Value</u>: Information possesses several dimensions of value, including as a commodity, as a means of education, as a means to influence, and as a means of negotiating and understanding the world. Legal and socioeconomic interests influence information production and dissemination.

   The value of information is manifested in various contexts, including publishing practices, access to information, the commodification of personal information, and intellectual property laws. The novice learner may struggle to understand the diverse values of information in an environment where "free" information and related services are plentiful and the concept of intellectual property is first encountered through rules of citation or warnings about plagiarism and copyright law. As creators and users of information, experts understand their rights and responsibilities when participating in a community of scholarship. Experts understand that value may be wielded by powerful interests in ways that marginalize certain voices. However, value may also be leveraged by individuals and organizations to effect change and for civic, economic, social, or personal gains. Experts also understand that the individual is responsible for making deliberate and informed choices about when to comply with and when to contest current legal and socioeconomic practices concerning the value of information.

4. <u>Research as Inquiry</u>: Research is iterative and depends upon asking increasingly complex or new questions whose answers in turn develop additional questions or lines of inquiry in any field.

   Experts see inquiry as a process that focuses on problems or questions in a discipline or between disciplines that are open or unresolved. Experts recognize the collaborative effort within a discipline to extend the knowledge in that field. Many times, this process includes points of disagreement where debate and dialogue work to deepen the conversations around knowledge. This process of inquiry extends beyond the academic world to the community at large, and the process of inquiry may focus upon personal, professional, or societal needs. The spectrum of inquiry ranges from asking simple questions that depend upon basic recapitulation of knowledge to increasingly sophisticated abilities to refine research questions, use more advanced research methods, and explore more diverse disciplinary perspectives. Novice learners acquire strategic perspectives on inquiry and a greater repertoire of investigative methods.

5. <u>Scholarship as Conversation</u>: Communities of scholars, researchers, or professionals engage in sustained discourse with new insights and discoveries occurring over time as a result of varied perspectives and interpretations.

   Research in scholarly and professional fields is a discursive practice in which ideas are formulated, debated, and weighed against one another over extended periods of time. Instead of seeking discrete answers to complex problems, experts understand that a given issue may be characterized by several competing perspectives as part of an ongoing conversation in which information users and creators come together and negotiate meaning. Experts understand that, while some topics have established answers through this process, a query may not have a single uncontested answer. Experts are therefore inclined to seek out many perspectives, not merely the ones with which they are familiar. These perspectives might be in their own discipline or profession or may be in other fields. While novice learners and experts at all levels can take part in the conversation, established power and authority structures may influence their ability to participate and can privilege certain voices and information. Developing familiarity with the sources of evidence, methods, and modes of discourse in the field assists novice learners to enter the conversation. New forms of scholarly and research conversations provide more avenues in which a wide variety of individuals may have a voice in the conversation. Providing attribution to relevant previous research is also an obligation of participation in the conversation. It enables the conversation to move forward and strengthens one's voice in the conversation.

6. Searching as Strategic Exploration: Searching for information is often nonlinear and iterative, requiring the evaluation of a range of information sources and the mental flexibility to pursue alternate avenues as new understanding develops.

    The act of searching often begins with a question that directs the act of finding needed information. Encompassing inquiry, discovery, and serendipity, searching identifies both possible relevant sources as well as the means to access those sources. Experts realize that information searching is a contextualized, complex experience that affects, and is affected by, the cognitive, affective, and social dimensions of the searcher. Novice learners may search a limited set of resources, while experts may search more broadly and deeply to determine the most appropriate information within the project scope. Likewise, novice learners tend to use few search strategies, while experts select from various search strategies, depending on the sources, scope, and context of the information need.

There is much to critique here (and, again I recommend reading Wilkinson's criticism). Also, since the Frames are designed to teach information literacy per se, I have not found all equally useful in thinking about how I want undergraduates to learn and practice a twenty-first-century digital source criticism in ancient studies. I will refer back the Frames as they are implicated in the assignments below, which are designed to impart specific information literacy lessons.

III. Some ideas for teaching critical information literacy in ancient studies

I tend to create three types of assignments with information literacy objectives. The first category includes assignments that ask students to use and then deconstruct digital models, in order to identify and analyze precisely the kinds of information that went into making them, often in comparison to a modern analog. The second category comprises activities that invite students to become active participants in the creation of information, as a way of encountering first-hand the impact of participation on the kind, quality, and amount of information in certain kinds of digital corpora and resources. Assignments of the third type ask students to step self-consciously out of their digital present and to recreate or solve information problems as an ancient person might have. The third category thus represents a sort of exercise in ancient information literacy, in order to cast into higher relief what is different and distinctive about our current information ecosystem. In this section I will give one example of each type of activity.

Type I: The critical use of digital models

If you teach Roman history, you may have come across or even taught with ORBIS, Stanford University's geospatial network of the Roman world.[15] ORBIS is a model of travel and connectivity in the Roman Empire that is capable of plotting various routes between any two of 632 sites, whose coordinates are taken from the online gazetteer Pleiades.[16] The routes mapped depend on certain key factors or constraints, such as the time of year (month or season), travel priority (the fastest, cheapest, or shortest route), travel medium (land, river, coastal, or open sea), and mode of travel (on foot, donkey, carriage; civilian or military; etc.). In addition to the routes, the model will also calculate the distance, time, and cost of the journey for a passenger and a kilogram of wheat, which allows for comparison of travel times and shipping costs at different times of year, according to different priorities, and along different routes. Finally, ORBIS is capable of mapping and comparing geospatial networks around a given central place in cartograms that represent the zones or isobars of distance as a function of time or cost. So, for instance, in a cartogram with Tarraco (mod. Tarragona) as the center point, Corinth and Corduba are represented as the same visual distance apart, and thus in the same functional zone, as each is calculated as being 14-15 days away in summer, despite the fact Corinth is much further away by geographic distance.

---

[15] http://orbis.stanford.edu/. ORBIS was reviewed recently by Chiara Palladino (2019).
[16] https://pleiades.stoa.org/

There is more that one can do with ORBIS, such as comparing multiple trade networks and exploring the effects of particular routes by excluding specific nodes; but for undergraduate teaching purposes, the functionality described above is particularly effective in demonstrating the likely nodal character and certain seasonality of connectivity in the ancient Mediterranean world, as well as the dramatic differences in time and cost between land and sea travel. In any ancient history or civilization course that touches on the Roman world, I typically spend part of a class or lecture modelling different routes to demonstrate these points, constructing at least a cartogram or two. Depending on the course, I also assign a break-out activity around ORBIS for a section (often led by graduate students) or turn this activity into a stand-alone paper topic, based on the section instructions, which asks the student to use and then critique the tool in an explicitly comparative mode.

The work of the section is divided into preparatory work to be done before class (the results of which I ask to be posted online the night before) and a set of operations and questions we try to perform and answer in class. (The instructions printed below are wordier than I tend to publish online, since I am including many questions here that I usually ask in person.) In order to facilitate in-class discussion, I organize the students into working groups of three to four people and assign to each group two primary sources from their sourcebook or textbook (translated inscriptions, papyrus documents, letters, excerpts from literary texts, etc.). I ask each student to prepare for the section by reading some of the online documentation for ORBIS and creating one travel scenario from a primary source. They are required to: describe the scenario; use it to model a route with ORBIS; and record and post the scenario and route online before class. The primary sources (e.g., Lewis & Reinhold (1990) *Roman Civilization*[3], Vol. 2, nos. 27, 28, or 30) all describe travel that is either germane to the subject matter of the class (e.g., the route between Rome and Alexandria) and/or include significant open sea (e.g., London to somewhere in the Mediterranean) or overland (e.g., anywhere in central Hispania to the Mediterranean) travel segments, since both kinds of trips will result in substantial differences in route, time, and cost if certain parameters are changed. I figure that student preparation takes approximately 30-45 minutes, if done conscientiously.

ORBIS Section instructions:

Before class:

1. Watch the three ORBIS (http://orbis.stanford.edu/) YouTube demonstrations in the "Using" tab in the "About" section.
2. Read the "Understanding," "Building," and "Geospatial" tabs in the About section. (If you have time, I also recommend reading Walter Scheidel's "Orbis: the Stanford geospatial network model of the Roman world" (http://orbis.stanford.edu/assets/Scheidel_64.pdf) and Scott Arcenas's "ORBIS and the Sea: a model for maritime transportation under the Roman Empire" (http://orbis.stanford.edu/assets/Arcenas_ORBISandSea.pdf), both of which are pdfs linked to the "Research" tab). Do not worry if you do not understand everything in the second and third tabs: please try to read them in light of the "Understanding" tab, which describes what one can expect of this model and why.
3. Create one specific travel or trade scenario based on one of the ancient primary sources you have been assigned. Describe the scenario you have constructed in two to three sentences, and try to be as specific as you can: Who is traveling and why? What is the origin and the destination? At what time of year are they travelling? Are they making any stops according to the source? What, if anything, are they shipping? Do we know anything about mode of transport? Etc. YOU WILL NEED THIS SCENARIO FOR CLASS.
4. Model the route for your scenario in ORBIS. In order to calculate the route, you will need to pick a set of characteristics, such as time of year, mode of travel, etc. Justify (i.e., give the reasons for) your settings as either most likely or based on something specific in your source material. Record the nodes (sites) of your route and the mode of travel, time, and cost for each leg of each journey.

ORBIS saves your searches in your search history. I recommend looking at this, so that you see how to toggle between searches in class. You can also print images of your routes. YOU WILL NEED THIS ROUTE FOR CLASS.
5. Post your scenario and route to the online discussion forum by 9pm the day before class.

In class:

6. Share your scenarios and routes with your group. Were they the same? How did they differ? Decide as a group on a final version of one scenario for each ancient source and model them in ORBIS, i.e., you need as a group to have two shared scenarios and routes based on your primary sources. Be sure to describe your final scenarios in two to three sentences and record the results of your routes (a good idea is to elect an official recorder for the group).
7. As a group, decide on at least one factor in each trip to modify: time of year (e.g., summer to winter); priority (e.g., cheapest to fastest); network modes (e.g., disallow travel by open sea, forcing the trip to go along the coast); or mode of travel (e.g., from foot to rapid military march for the land leg and from civilian to military for river travel). Recalculate and re-describe the routes. (Again, the recorder should make sure that you have notes for the results of your new routes.) Are they different? How? What accounts for the differences? For instance, what would happen if you were to take the same journey only by road? How much does the season matter and what is affected? How important are rivers to your route with respect to time or cost? Post your results online to the class forum.
8. Go back to your ancient source: Did ORBIS map the same itinerary as what seems to be described? If not, what is different? Can you think of reasons why? Is there even enough information in your source to know what the itinerary was? (For thinking about these and the following questions, I recommend reviewing the "Understanding" and "Building" tabs in the About section.)

9. What kind of information does ORBIS model? Where does it come from? *When* does it come from? Consider, for example, the information used to calculate prices: what is the source for that? What are some of the pros and cons of relying on this source for the purpose of this model? To what extent does ORBIS seem to rely on sources like the ones from which you derived your scenarios? What sort of information do you think it takes from those kinds of sources and how does it seem to incorporate it? Do the answers to these questions have implications for what this model is telling us when it calculates a route? How "Roman" is this model? How "imperial"? Can we use this model to think about the Mediterranean ca. 400 BCE? How about ca. 800 CE?

10. What does ORBIS leave out? In other words, are there factors, which were likely important to the cost and duration of any ancient trip, that the model does not include? Can you see any of these factors implicated in the specific scenario you modeled? In thinking about this question, it might be useful to try to retrace the steps in any long, multi-leg journey you have taken and consider the factors that made that trip deviate from some notional "average."

11. When ORBIS was first introduced, several journalists in the popular press called it a Google Maps for ancient Rome (examples are collected in the "Media" tab in the About section). One can see why they made this comparison, but is it apt? Why or why not? Are the similarities or differences between ORBIS and Google Maps more important?

    a. Now that you have thought about the kind, quality, and amount of data that OBRIS is integrating when it calculates a "route" with associated times and costs, we need to explore how Google Maps works. There are several popular descriptions of how Google Maps works, but the most useful summary I know, with links to many of those resources, is the article in Wikipedia.[17] Many of the technical details are complicated, but please see if you can figure out some of the

---

[17] https://en.wikipedia.org/wiki/Google_Maps#Map_data_and_imagery

kinds of information Google draws on and how much. What kind of data does it collect in order to calculate travel times? How does it compare to that which is collected and compiled for ORBIS with respect to type, quality, and amount?
- b. Is the primary aim of Google Maps to describe or predict? What about ORBIS? If you see a difference in aim, is this important? You probably look to Google Maps to give you a useful answer to a precise travel question: does ORBIS provide the same sorts of "answers" to the same sorts of questions? Are Google Maps results and ORBIS results "useful" in the same way, or do we use the results differently? What are we supposed to "do" with an ORBIS result?
- c. Taking into account what you now know about Google Maps, what are the key similarities between it and ORBIS? What are some of the key differences? Are the differences quantitative or qualitative or both? Which are more important, the similarities or the differences? In your opinion, is it helpful to say that ORBIS is a Google Maps for the Roman world? Why or why not?

12. In the final analysis, what does ORBIS tell us about travel and connectivity in the ancient world? On another level, what does ORBIS tell us about our ability to build sophisticated digital models of the ancient world? Do the differences in kind, quality, and quantity of information available to us now mean that we have a fundamentally different relationship to antiquity than to the present and recent past? If so, do you think that the ORBIS interface should make this difference clear? For instance, you now likely have a much deeper appreciation of the limits of ORBIS: should the interface or the results give some sort of obvious indication of those limits as a warning or reminder to the user?

In class, I and/or the graduate student(s) move from group to group, asking questions and driving them forward or throwing a provocative monkey wrench into the works, as required. At a certain point, perhaps 20-25 minutes into the period, I bring the groups together to discuss what they have discovered by doing steps 6-10. I often ask one group to present its scenarios, routes, and transformations, which (hopefully!) have been posted online, to serve as a focal point for conversation. We then address some of the questions raised in steps 8-10. This leaves 20 minutes or so to explore and discuss steps 11-12. We return the students to their groups and charge them with staking out a position on the comparison of ORBIS to Google Maps. I give them about 10 minutes to organize their positions and then we reconvene to discuss. The essay version of this activity is almost like a lab report: the student constructs a scenario or two from a primary source and then maps the routes and transforms them; she argues what she believes these experiments with the model reveal about travel in the Roman world; she finally compares ORBIS to Google Maps with a view to how we are to understand and use ORBIS results as evidence for travel in the Roman world.

In the age of black-box devices and seamless apps, I have found it increasingly important and useful to have the students meditate on what one might view as twenty-first-century digital source criticism, since the majority of our digital models of antiquity are not built on the same kind, quality, or quantity of data as those which constitute the main points of departure and reference for our students. When crafting this sort of assignment with an information literacy objective, I tend to go back to the Framework and the associated practices and dispositions as a stimulus to thinking about the kinds of questions I want to ask the students.

Frame 4 (research as inquiry) is integral to the design of the session above, since research as inquiry is built into the DNA of the ORBIS model: one of the main points in creating the scenarios and then working the transformations is not so much to learn how to *use* the tool (e.g., how to retrieve the text of an inscription in a more traditional database, like the Clauss-Slaby epigraphic database),[18] much less to discover "the" route between A and B (which route, as Scheidel notes, would be completely coincidental to that of any recorded ancient trip), but rather to explore the *heuristic value* of a model like ORBIS by playing with the parameters and measuring

---

[18] http://www.manfredclauss.de/

the results against other forms of evidence. In other words, it is only by asking questions of the model that we succeed in unearthing and interrogating our own assumptions and discovering and testing new patterns latent in our data.

Similarly, Frame 1 (authority and expertise) is useful for thinking about how I want students to learn to see the subjectivity encoded in the "data" that underlies a model like ORBIS. Scheidel is both an expert and an authority, and he has done an excellent job in making his assumptions and choices clear in the documentation to ORBIS; but one can easily imagine that another editor might have made different decisions, with a potentially profound impact on the results. Significantly, students develop over the course of this session a much deeper appreciation of why the citation to any map one prints from ORBIS generates a citation with Scheidel and Meeks as the "authors." They come to see map they have created not as one tracing "the" route from, e.g., Rome to Sirmium, but rather as one illustrating an outcome of Scheidel's and Meeks's hypothesis about how travel worked in the Roman Mediterranean. (I sometimes bookend class by asking about this citation, to see how their views change from start to finish.) This sort of observation feeds into Frame 5 (scholarship as conversation), as we come to realize that ORBIS is in fact more of a planting of an intellectual flag in the field of scholarly research on the Roman world than a "tool" to "answer" a question.

In many ways, this and other exercises of this type are really extended meditations on Frames 2 and 6 (information creation as process and searching as strategic exploration). The main task of this section, for which ORBIS is a case study, is to think as precisely and explicitly as we can about the effects of taking the evidence we have for the ancient world, like the isolated testimonials for travel in ancient literature and documents underlying our scenarios, and reprocessing and repackaging that evidence as visual representations which are perhaps best described as bits of fact stitched together with a relatively large number of (reasonable, but debatable) inferences drawn from comparative sources. I have found that there is something very powerful in pointing out to students that over the course of this exercise we have moved from a traditional, nineteenth- and twentieth-century medium and mode for studying ancient history (i.e., the source book, which presents a subjective and tendentiously edited collection of translated sources, not the evidentiary *Dinge an sich* in all of their messy, unmediated reality) to a twenty-first-century medium and mode (which has its own set of epistemological problems). The only way one can truly begin to comprehend the intellectual value of ORBIS results as "evidence" or "information," and so be strategic in their generation and deployment in an argument, is to understand the process by which these visualizations are constructed.

All of these critical aims are thrown into high relief by comparing ORBIS results to those one gets from what seems to be a modern analog, Google Maps. Besides the obvious differences in aim as well as sources, quality, and amounts of information (the data underlying Google Maps is: collected systematically and from multiple sources; generated and shared as structured data; represents actual routes and actual trips based on crowdsourced GPS and real-time accident reporting data—none of which one can claim for any travel data from the ancient world), there are some other salient points of divergence. For instance, where is the documentation behind Google Maps? There is very little, because information has (commercial) value (Frame 3). Also, you might say that Google Maps has a very different authority problem from that underlying ORBIS, because the proof is in the pudding: it either gets you where you want to go, when it predicts it will, or it does not. It is precisely the uncritical and implicit transference of this kind of authority structure, erected for our modern predictive digital models, that this lesson aims to expose and deconstruct for our descriptive digital models of the ancient world.

Type II: Community-based digital resources

As described above, one of the principal aims motivating the drafting of the Framework was to open up a space for the social, participatory dimension of the information environment in which we now find ourselves. The Standards were published in 2000 and Wikipedia was launched in January of 2001: obviously, a good deal had changed between 2000 and 2015. In one sense, of course, scholarship has been "social" and "participatory" for more than a century, or at least one could make that argument when looking back on the growth of journals and international associations since the turn of the twentieth century or the organization of massive, collaborative undertakings like the Pauly-Wissowa, the *Corpus Inscriptionum Latinarum*, the *Lexicon Iconographicum Mythologiae Classicae*, etc. Even crowdsourcing is not itself a new concept: the "premium edition" of a book in the early nineteenth century was one that had been edited by the public with a reward paid for each error discovered in the proof-sheets; on a much more ambitious scale, the *Oxford English Dictionary* was effectively crowdsourced, as vividly related in Simon Winchester's *Professor and the Madman* (1998).[19] Since 2000, however, changes in the technology, scale, and application of the social, participatory mechanics of collaboration have transformed the speed and modalities of scholarly debate, communications, and cooperation, often in ways that are not always obvious to students. Understanding the implications of the social and participatory elements of current scholarship goes beyond the initial and often overly narrow focus on the admittedly important issues of authority, credibility, and perspective. Some of the most interesting and useful resources and corpora or repositories in ancient studies today are the product of participatory projects, such as Pleiades, papyri.info, the Online Coins of the Roman Empire (OCRE), the Nomisma.org project, and Open Latin and Greek (OGL), to name a few that I personally use for research and teaching; and, like Wikipedia, one cannot use them critically without understanding how the data gets there, who is allowed to transform it and how, and what its limitations are. Activities and assignments under this rubric thus aim to push students to understand something of the mechanics, rules, limits, and ethics of scholarly participation, and how each affects the shape and quality of the information they retrieve from these digital resources. Below I give some of the exercises I assign with papyri.info in various types of courses.

---

[19] Crowdsourcing was a technique understood by the Greeks: see, e.g., Arist. *Pol.* 1281a40-b10. For Greek crowdsourcing in practice, see, e.g., Lanni 2016: Ch. 2.

Papyri.info and crowdsourced scholarship

Papyri.info is a web application that aggregates and allows searching via the Papyrological Navigator (PN) of several different kinds information from a collection of increasingly integrated databases, including the Advanced Papyrological Information System (APIS, which consists of metadata records edited by the institution holding the ancient texts), the Duke Databank of Documentary Papyri (DDbDP, which originally was dedicated to collecting and encoding the Greek and Latin texts of ancient papyrological documents), the Heidelberger Gesamtverzeichnis der griechischen Papyrusurkunden Ägyptens (HGV, which created metadata records for ancient documents for information such as date, provenance, publication, keywords, etc.), and the Bibliographie Papyrologique (BP, which collects and publishes bibliography on papyrological subjects). Papyri.info also serves as a portal for the Papyrological Editor, an online text editor that allows registered users to enter, edit, and (if they have the requisite editorial privileges) approve digital versions of papyrological texts in TEI EpiDoc XML.

Whenever I teach ancient history or culture classes in translation, I always attempt to make time during a lecture in which papyri figure prominently (typically, classes on the ancient economy, family relations, literacy, government, etc.) to step strategically out of the lecture and make—in real time—a simple addition or correction to a record in papyri.info. This may seem like a distraction from the topic at hand, but I find that it provides good pedagogical value: signing into papyri.info, transforming a record (and here it does not matter if the correction is to the Greek or Latin, the translation, the punctuation, the bibliographic metadata, etc.), committing the change to the editorial boards, and pointing out how such work is recorded, takes perhaps eight to ten minutes at most (I abandon ship if the site happens to be slow); yet those eight to ten minutes succeed admirably in lifting the hood on an important scholarly digital resource. The majority of my students are completely removed from the core scholarly activities of reading ancient sources in the original languages and formats, excavating on site, handling artefacts, etc. For this reason, they not only find it interesting to be invited into the scholar's workshop, but they also learn some valuable lessons as to what this process means for the information presented on papyri.info and similar digital resources for antiquity (and beyond).

For example, the students see that the structure of papyri.info accommodates two modes or levels of participation: one for capturing and sharing volunteer contributions; and one for exercising a form of expert peer review, since all changes must be reviewed and approved by editors before being pushed to the public (in this light, the fact that my change will not go through instantaneously is part of the lesson). This observation affords us a chance to discuss authority and expertise and the mediation of the data presented in papyri.info (cf. Frame 1). After I submit my contribution, I point out the documentation of past transformations that is attached to each record and my own personal record of microattributions for scholarly interventions (e.g., offering a textual emendation or supplement, correcting a reading from the original, etc.) and scholarly service (e.g., adding a text from an *editio princeps*, correcting miscoded lineation or punctuation, etc.). I also show them how some changes (and not others) are collected and displayed with the text or in the *apparatus criticus*, and (in two slides) how all of this is replacing (but has not yet fully replaced) the twentieth-century scholarly tools of the *Sammelbuch* (which collects and republishes in print a corpus of all papyrological editions, assigning to each a unique publication number) and the *Berichtigungsliste* (which collects, collates, and republishes editorial and scholarly corrections to published papyri).[20] In terms of the Framework, we here see scholars actively engaged with the idea that information has value, as demonstrated by the care that they have taken to properly record and credit all scholarly work (Frame 3).

Tracking who has done what to each text also allows us to think about these texts as (hierarchical, structured) scholarly "conversations"—if that really is the right metaphor (Frame 5).[21] In order to help illustrate the contours of the papyrological conversation, I show how the majority of texts have in fact been edited by a small number

---

[20] Preisigke et al. eds. 1915- and 1913-, respectively.
[21] Cf. Wilkinson 2014b and 2016e.

of editors (which is conveniently visualized by Trismegistos Editors) and suggest (admittedly, in a more anecdotal way) via the activity statistics available in the editorial interface that the same is likely true of the number of active contributors to papyri.info versus the number of users.[22] Papyri.info allows us to see the evolution of the texts, and so the "conversations" that they embody, through time. Indeed, the evolutionary, open-ended, potentially unfinished character of the texts is one of the main information literacy lessons I hope to communicate to the students. More specifically, I argue that the current state of the texts in papyri.info reflects two important drivers in the scholarly ecosystem: the present state of our scientific knowledge; and the ability, time, and commitment of a discrete community of scholars to contribute its time and encode its knowledge in shared XML records. In other words, one cannot assume that any record is either correct (indeed, did I not just now correct an error, albeit a relatively minor one?) or current, or that the database as a whole reflects the entire universe of published texts. In other words, one must always ask: Does the text here reflect a decades-old *editio princeps* or our most current reading? Has anyone had the time to incorporate all the corrections of the *Berichtigungsliste*? How about any or all of the corrections and advances of the last two years? Are there texts important to my search that have been published recently but not yet encoded and so will not show up in my search results? And so on.

Most undergraduates taking ancient studies classes in translation are unlikely to need to work with the texts in papyri.info so closely as to make the kind of critical window I offer above directly useful. But again, that is not the point of this ten-minute exercise: instead, the aim is to get them to see how the scholarly sausage is made (in a way that is not possible with a closed model like ORBIS, a direct comparison I draw if I have taught both) and to prod them to consider how that process implicates structures of expertise and authority and the credibility and quality of the information in a contemporary scientific corpus. In undergraduate Latin and Greek classes, I plan a more

---

[22] Trismegistos Editors: https://www.trismegistos.org/edit/index.php. This long-tail phenomenon appears to be generally true of wiki-type projects, including Wikipedia: see Matei and Britt 2017.

hands-on version of this type of exercise. For example, I assign a papyrus letter to translate in class alongside a literary letter by, say, Cicero or from Chariton's novel *Chaereas and Callirhoe*. I follow this up with a written assignment in which I have them pick a papyrus letter that has no online translation from a list I have assembled and ask them to contribute their own translation to papyri.info. To do this, they need to sign up for an account (which is very easy) and I teach them the rudiments of textual markup and how to contribute a translation via the text editor, with the help of the online documentation available. I work with them individually on the translations, to which they append a short commentary in which they make explicit and justify their philological choices (this commentary is for class purposes only: it does not get uploaded to papyri.info). I have found that students take this assignment very seriously and appreciate the opportunity to join the scholarly conversation. It also expands their view of what should now count as "publication" and certainly gives them a deeper appreciation of the costs imposed on both a community and individual level when it comes to maintaining and growing a "free" corpus of ancient texts (cf. Frame 3). The recent development of the Digital Corpus of Literary Papyri (DCLP) and the Digital Latin Library (DLL) now means that there are an increasing number of similar exercises one could design using literary and subliterary papyri and texts.[23]

Type III: Ancient information literacy

One of the lessons I hope to inculcate in assignments of Type III is a recognition of the fact that the data we have for the ancient world differs not only quantitatively but also qualitatively from what we have for the contemporary world.[24] While the former is obviously in large part a function of survival, the latter is a function of the measurement and data habits in antiquity: even if all documents had survived, we would still be missing much of the information we should like to have, since it was not recorded in the first place. This negative lesson,

---

[23] DCLP: http://www.litpap.info/; DLL: https://digitallatin.org/
[24] Cf. Dunn 2012 on the terms "qualitative" and "quantitative" data in the humanities.

in my experience, is one worth teaching students, particularly in this context, because it subverts many of their implicit assumptions and helps them to become more critically aware of the contours of their own modern relationship to information and data. In other words, it helps to reveal our own structures and techniques of measuring, cataloging, indexing, discovering, accessing, authenticating, and communicating to ask of ancient people the sorts of questions the Framework presses our students to consider in their own lives:

- What sort of information did ancient people collect, and why and how?
- How did they store, discover, retrieve, present, share, and guarantee the integrity of information they used, all without computers? Furthermore, how did these capacities, techniques, and modalities affect teaching, research, planning, and dispute resolution (to name but a few core social activities)?
- Who or what did ancient people trust and why?
- What techniques of authentication did they devise?
- How were any of these concepts or techniques taught?
- Conversely, how did people take advantage of the systems they built for their own ends? For example, was ancient information ever "stolen? Were there fakes, forgeries, disinformation, and "fake news" in the ancient Mediterranean world?

I have found Josh Ober's *Democracy and Knowledge* (2008) useful in thinking about the relationship of politics, culture, and information in classical Athens, and I am confident that Andrew Riggsby's recent monograph, *Mosaics of Knowledge* (2019), will provoke an interesting discussion of similar questions for the Roman Empire.[25] Below are two paper topics I assign in order to encourage students to think about information literacy in the ancient Mediterranean world, and by extension information literacy in their own. The first topic

---

[25] Johnstone 2011 is also interesting from this perspective.

comes from a seminar in translation for non-classics majors which explored institutions, economics, information, and strategic behavior in the Greek world. The second was assigned in a general education course on Greco-Roman Egypt.

- *Multifactorial authentication in the Hellenistic world?*
  Analyze the inscription recorded from Hellenistic Paros and published as *SEG* 33.679.[26] Paros was a polis on the eponymous island in the Cyclades. The inscription reports laws passed as a reform in the wake of a scandal at the *mnēmoneion*, a public record office where people could deposit notarized copies of their business documents. What was the nature of the scandal? What reforms did the Parians implement? What effect will these reforms have on the transaction costs of doing business in Paros? Do you see any familiarities between the problems confronting the Parians or the solutions they adopted and the experience of other communities in Classical Greece? As you reconstruct the problem and the solution embodied in this inscription, you may wish to think about the roles of literacy, documents, archives, law, inscriptions, enforcement, expertise, etc. You may also wish to compare this problem and its proposed solution to current issues surrounding fraud, authenticity, and information and identity control.

- *Information and the Ptolemaic State.*
  Using at least 2-3 documents from Bagnall and Derow, *The Hellenistic Period* (2004), make an argument about the role of information in the management of Ptolemaic Egypt. In thinking about this topic, please be sure to ask yourself what sort of information you are talking about (be specific: try

---

[26] A German translation of the inscription is available in Lamnbrinoudakis and Worle 1983. I have found published English translations of this interesting inscription deficient. I therefore distribute my own English translation with notes, which I am happy to share upon request. For two useful recent overviews of archives and information management in classical Greece, see Faraguna 2015 and Harris 2015.

focus on just one or two types of information) and how it was generated, compiled, accessed, authenticated, and shared. You might also want to look for ways in which various parties exploited these types information and the structures meant to control it for their own ends. Good documents for this topic are: B&D 84, 86, 87, 89, 90, 92-95, 99, 100, 102, 103, 105, 106, 107, 110, 114, 116, 117, and 124.

Conclusion

Lisl Walsh in her contribution argues eloquently for what we might call a threshold concept in Classics, which for her encompasses an integrated methodological approach to the ancient Mediterranean, not just philology: "partial-puzzle analytics." By this she denotes an intellectual approach, which she sees as specific to Classics, that combines rigorous micro-analysis with creative but controlled extrapolation and contextualization of limited evidence into a larger picture, whose outlines are barely adumbrated. In other words, to "think like a Classicist" is to learn to scrutinize closely the few remaining pieces of a large and complex puzzle and to put those surviving pieces in their proper places without the benefit of the picture on the cover of the box. (Paleontology and evolutionary biology would seem to depend on a similar set of skills.) For Walsh, the addition of the "digital" is compatible with teaching Classics and the inculcation of the deep learning of partial-puzzle analytics, but not essential. In fact, she sees a latent but potentially existential risk in digital approaches. First, in her view many digital projects and techniques, which aim to smooth or extrapolate from limited data, often therefore work to obscure the heterogeneity, distribution, and essential gapiness of the underlying evidence, which for her is the foundational methodological point of departure for Classics as a distinct and distinctly valuable intellectual discipline. Second, she also worries that teaching digital techniques, often in response to pressure to make the humanities "relevant," can effectively supplant, rather than support, the teaching of Classics.

I find Walsh's vision of Classics as an exercise in partial-puzzle analytics deeply compelling. I also share her wariness (echoed by Caraher in this collection) of the uncritical use of digital approaches in ancient studies classrooms, often driven by what I see as a grimly myopic and ultimately self-defeating techno-philistinism. However, I have drawn the opposite conclusion, finding myself instead motivated to think about how to introduce digital and information literacy approaches into my teaching precisely because we are living in a world dynamically shaped by digital approaches. In this contribution, I argue that teachers of the ancient Mediterranean should consider incorporating some information literacy lessons into their curriculum for (at least) two reasons, both of which are, for me, deeply implicated in the teaching of partial-puzzle analytics.

First, our students come to us at least information semi-literate, but crucially in a different information culture. When they approach digital resources—and their number will only increase—they are likely to draw many of the inferences that so concern Walsh: that the data is complete, objective, standardized, etc. When they use ORBIS, they see Google Maps. I have found ORBIS to be a valuable didactic and research tool for visualizing certain important aspects of travel in the ancient world, but it is, in essence, an interactive manifestation of one team's intellectual picture of one part of the ancient Mediterranean puzzle. From this perspective, it is hard to think of a better way to demonstrate the essential partial-puzzle reconstructiveness of scholarship on the ancient world than to deconstruct so seemingly complete a reconstruction as ORBIS. That said, to my mind one of the most promising frontiers in digital resources is not exemplified by closed digital projects like ORBIS, but by open, community-based, collaborative projects, like some those I note under Type II. From where I stand, with one foot in libraries and the other in the research community, open, community-based resources are poised to play an increasingly important role in the ancient studies research ecosystem. We who use—and perhaps particularly those of us who contribute to—these resources have a positive pedagogical obligation to teach students about the ways in which such resources are built and sustained, and how this affects the information they contain. My first suggestion, then, is that we should teach information literacy about digital approaches and resources as a form of twenty-first-century source criticism in ancient Mediterranean studies.

Second, the low-information environment of the ancient Mediterranean world is, for me, one of the important markers of its pre-modernity. It is not merely that much of the documentation and information has perished with time (although this is true), but also that the ancient world was radically and perhaps essentially unmeasured compared with our modern society. We tend to see the effects of both as "information gaps," but they should not be conflated, since they represent distinct phenomena, with the latter having, I would argue, a profound effect on the ancient experience. Teaching in an information literacy mode, precisely because it was conceived to deal with the complexities of negotiating the information age, can help to delineate both the gaps of survival and the contours of an increasingly alien information culture.[27] For instance, keyword searching in the TLG or any other digital literary corpus is self-evidently useful; but does it matter that we can now read with a completeness and a precision that no Callimachus or Horace could have ever contemplated, much less attempted? How did they or the Aristarchuses or Galens of antiquity search or compare literary texts? To be sure, certain words seem to be keywords in ancient poetics or political discourse, but how precisely did they act as "keywords" if keyword searching was effectively impossible? Seen in this light, what do our search results *mean* when comes to, say, the actual practices of *ancient* intertextuality?[28]

---

[27] Cf. Riggsby 2019: 2.

[28] Fowler's essay (1997) on the meaning (in all senses and directions) of intertextuality remains a classic: my question is specifically about the relationship of *ancient* searching to *ancient* intertextuality. In other words, how do we imagine ancient authors, readers, and critics went about the sort of operations Fowler describes for himself using the PHI corpus of electronic texts on pp. 20-24. And further, is there any evidnece of the ways in which the knowledge of those ancient reading and searching strategies and techniques conditioned the writing or interpretation of texts in antiquity. With Fowler (31), should we be open and sensitive to the possibility that the potential and quality of ancient intertextuality evolved from the 5th century BCE to the 5th century CE, as the number and availability of texts increased? For a recent description of current forms and trends in intertextual searching, see Coffee 2018. Coffee outlines four scenarios of modern intertextual practice or operations, but only the first was possible in antiquity, since it begins and ends with human reading and memory. The other three involve targeted or computational searching of texts, corpora, and tagged and encoded intertexts. He contends that his fourth scenario, which envisions reading with a

Similar questions can and should be asked of ancient politics and administration on the basis of our surviving documentary record, which is an archaeology of ancient information and information practices. To ask students to try to reconstruct ancient information techniques and strategies from what survives is thus to ask them to step out of one of the key ways in which they are most self-consciously "modern" and to inhabit temporarily a world characterized by the particular limits, freedoms, thought-patterns, and ingenuities of a comparatively well-documented pre-digital age.

---

visual, customizable, instantaneously available, linked, and shareable web of texts, translations, and intertexts, "could simulate the experience of the ancient one" (220). This *might* approximate or recover something of the otherwise lost mental condition of the highly educated, urban, élite reader who had consumed a steady diet of Latin and Greek texts from an early age and had access to an exceptional library. If so, we might see this as our reading with a sort of ancient reading or memory prosthesis. But to my mind, this sort of reading more likely misses or obscures what was essential to the condition of most ancient reading, and so the precondition of ancient text production, namely that texts were hard to find and harder to search; that many texts or discourses were oral and visual and local; and that intertexts were themselves hard to find or to share when found because citation was rudimentary and non-uniform. This is not to say that new ways of reading are not valuable or do not recover some important ways of ancient reading or intertexuality, only that we should mind the gap between the ancient and modern.

Works Cited

Framework for information literacy for higher education.
   2015. Retrieved August 26, 2019 from http://www.ala.org/acrl/files/issues/infolit/framework.pdf.

Information literacy competency standards for higher education.
   2000. Retrieved August 26, 2019 from http://www.acrl.org/ala/mgrps/divs/acrl/standards/standards.pdf.

Bagnall, Roger S. and Peter Derow
   2004 *The Hellenistic Period: Historical Sources in Translation. Blackwell Sourcebooks in Ancient History*. Blackwell, Malden, MA and Oxford. DOI: 10.1002/9780470752760

Bombaro, C.
   2016 The Framework is Elitist. *Reference Services Review* 44(4): 552-563. DOI: https://doi.org/10.1108/

Beyerle, S.
   2016 Authority and propaganda: The case of the Potter's Oracle. In *Sibyls, scriptures, and scrolls* edited by J. Baden, H. Najman, and E. J. C. Tigchelaar, pp. 167-184. Brill, Leiden. DOI: 10.1163/9789004324749_012.

Coffee, N.
   2018 An Agenda for the Study of Intertextuality. *Transactions of the American Philological Association* 148(1): 205-223. DOI: 10.1353/apa.2018.0008.

Collins, J. J.
   1994 The Sibyl and the Potter: Political propaganda in Ptolemaic Egypt. In *Religious propaganda and missionary competition in the New Testament world: Essays honoring Dieter Georgi* (Supplements to the Novum Testamentum 74), edited by L. Bormann, K. Del Tredici, and A. Standhartinger, pp. 57-69. Brill, Leiden. DOI: 10.1163/9789004267084_005.

Dawes, L.
   2019 Through faculty's eyes: Teaching threshold concepts and the Framework. *portal: Libraries and the Academy* 19: 127-53. DOI:10.1353/pla.2019.0007.

Dolinger, E.
2019 Defining and teaching information literacy: Engaging faculty and the Framework. *College & Research Libraries News* 80: 10-13, 21. DOI: 10.5860/crln.80.1.10.

Dunn, S.
2012 Review of ORBIS. *Journal of digital humanities* 1(3). Retrieved August 26, 2019 from http://journalofdigitalhumanities.org/1-3/review-of-orbis-project-by-stuart-dunn/.

Faraguna, M.
2015 Archives, documents, and legal practices in the Greek polis. In *The Oxford Handbook of Ancient Greek Law* edited by E. M. Harris and M. Canevaro. Oxford University Press, Oxford. DOI: 10.1093/oxfordhb/9780199599257.013.14.

Fish, S.
1980 What Makes an Interpretation Acceptable? In *Is there a text in this class? The authority of interpretive communities* edited by S. Fish, pp. 338-355. Harvard University Press, Cambridge, MA.

Foasberg, Nancy M.
2015 From Standards to Frameworks for IL: How the ACRL Framework Adresses Critiques of the Standards. *portal: Libraries and the Academy* 15(4): 699-717. DOI: 10.1353/pla.2015.0045

Fowler, D.
1997 On the shoulders of giants: Intertextuality and Classical Studies. *Materiali e discussioni per l'analisi dei testi classici* 39: 13-34.

Gruen, E. S.
2016 When is a revolt not a revolt? A case for contingency. In *Revolt and resistance in the ancient classical world and the Near East* edited by J. J. Collins and J. G. Manning, pp. 10-37. Brill, Leiden. DOI: 10.1163/9789004330184_003.

Harris, E. M.
2015 The legal foundations of economic growth in Ancient Greece. In *The Ancient Greek Economy: Markets, Households and City-states*, edited by E. M. Harris, D. M. Lewis, and M. Woolmer, pp. 116-146. Cambridge University Press, Cambridge. DOI: 10.1017/CBO9781139565530.006.

Johnstone, S.
 2011 *A History of trust in ancient Greece*. University of Chicago Press, Chicago.

Kerkeslager, A.
 1998 The Apology of the Potter: A translation of the Potter's Oracle. In *Jerusalem Studies in Egyptology (AAT* 40), edited by I. Shirun-Grumach, pp. 67-79. Harrassowitz, Wiesbaden.

Koenen, L.
 1968 Die Prophezeiungen des 'Töpfers'. *ZPE* 2: 178-209. Retrieved from http://www.jstor.org/stable/20180111.
 2002 "Die Apologie des Töpfers an König Amenophis oder das Töpferorakel." In *Apokalyptik und Ägypten. Eine kritische Analyse der relevanten Texte aus dem griechischrömischen* Ägypten (OLA 107, edited by A. Blasius and B. U. Schipper, pp. 139-187. Peeters, Leuven.

Ladynin, I.
 2016 Virtual history Egyptian style: The isolationist concept of the Potter's Oracle and its alternative. In *Greco-Egyptian interactions: Literature, translation, and culture, 500 BCE-300 CE*, edited by I. Rutherford, pp. 163-186. Oxford University Press, Oxford. DOI: 10.1093/acprof:oso/9780199656127.003.0007.

Lambrinoudakis, W. and M. Worle
 1983 Ein hellenistische Reformsgesetz über das offentliche Urkundenswesen von Paros. *Chiron* 13: 283-368. URN: nbn:de:0048-chiron-1983-13-p283-368-v5633.1.

Lanni, A.
 2016 *Law and order in ancient Athens*. Cambridge University Press, Cambridge. DOI: 10.1017/CBO9781139048194.

Latham, D., M. Gross, and H. Julien
 2019 Implementing the ACRL Framework: Reflections from the field. *College & Research Libraries* 80: 386-400. Available at: https://crl.acrl.org/index.php/crl/article/view/17397.

Leaning, M.
 2017 *Media and information literacy: An integrated approach for the 21st century*. Chandos, Cambridge, MA.

Lewis, N, and M. Reinhold
1990 *Roman Civilization: Selected Readings*. Third Edition. 2 Volumes. Columbia University Press, New York.

Ludlow, F. and J. G. Manning
2016 Revolts under the Ptolemies: A paleoclimatological perspective. In *Revolt and resistance in the ancient classical world and the Near East*, edited by J. J. Collins and J. G. Manning. pp. 154-171. Brill, Leiden. DOI: 10.1163/9789004330184_011.

Mackey, T. P. and T.E. Jacobson
2011 Reframing information literacy as a metaliteracy. *College & Research Libraries* 72(1): 62-78. DOI: https://doi.org/10.5860/crl-76r1.

2014 *Metaliteracy: Reinventing Information Literacy to Empower Learners*. Neal-Schuman, Chicago.

Mackey, T. P. and T.E. Jacobson, editors
2016 *Metaliteracy in practice*. Neal-Schuman, Chicago.

Matei, S. A. and B.C. Britt
2017 *Structural differentiation in social media: Adhocracy, entropy, and the "1% effect."* Springer, Cham, Switzerland. DOI: 10.1007/978-3-319-64425-7.

McGing, B.
2016 Revolting Subjects: Empires and Insurrection, Ancient and Modern In *Revolt and resistance in the ancient classical world and the Near East*, edited by J. J. Collins and J. G. Manning, pp. 139-153. Brill, Leiden. DOI: 10.1163/9789004330184_010.

Meyer, J.H.F. and R. Land
2003 Threshold concepts and troublesome knowledge: Linkages to ways of thinking and practicing." In *Improving student learning: Theory and practice ten years on*, edited by C. Rust, pp. 412-424. Oxford Centre for Staff and Learning Development (OCSLD), Oxford.

2005 Threshold concepts and troublesome knowledge (2): Epistemological considerations and a conceptual framework for teaching and learning. *Higher Education* 49(3): 373-88. DOI: 10.1007/sl0734-004-6779-5.

Meyer, J.H.F. and R. Land, editors
2006 *Overcoming barriers to student understanding: Threshold concepts and troublesome knowledge.* Routledge, London. DOI: 10.4324/9780203966273.

Oakleaf, M.
2014 A roadmap for assessing student learning using the new framework for information literacy for higher education. *The Journal of Academic Librarianship* 40(5): 510-14. DOI: 10.1016/j.acalib.2014.08.001.

Ober, J.
2008 *Democracy and knowledge: Innovation and learning in classical Athens.* Princeton University Press, Princeton, NJ.

Palladino, Ch.
2019 Review: ORBIS: The Stanford geospatial network model of the Roman world. *Society for Classical Studies Blog,* September 5, 2019. https://classicalstudies.org/scs-blog/chiara-palladino/review-orbis-stanford-geospatial-network-model-roman-world.

Potter, D. S.
1994 *Prophets and emperors: Human and divine authority from Augustus to Theodosius.* Harvard University Press, Cambridge, MA.

Preisigke, F. et al. editors
1913- *Berichtigungsliste der griechischen Papyrusurkunden aus Ägypten.* Vereinigung Wissenschaftlicher Verleger, Berlin.

Preisigke, F. et al. editors
1915-. *Sammelbuch griechischer Urkunden aus Ägypten.* K.J. Trübner, Strassburg.

Riggsby, A.
2019 *Mosaics of knowledge: Representing information in the Roman world.* Oxford University Press, Oxford. DOI: 10.1093/oso/9780190632502.001.0001.

Townsend, L., K. Brunetti, and A.R. Hofer
2011 Threshold Concepts and Information Literacy. *portal: Libraries and the Academy* 11: 853-69. DOI:10.1353/pla.2011.0030.

Wilkinson, L.
2014a *The problem with threshold concepts*. Sense & Reference, June 19, 2014. Retrieved August 26, 2019 from https://senseandreference.wordpress.com/2014/06/19/the-problem-with-threshold-concepts/.
2014b Is scholarship a conversation? *Sense & Reference*, July 15, 2014. Retrieved August 26, 2019 from https://senseandreference.wordpress.com/2014/07/10/is-scholarship-a-conversation/.
2014c Is research inquiry? *Sense & Reference*, July 15, 2014. Retrieved August 26, 2019 from https://senseandreference.wordpress.com/2014/07/15/is-research-inquiry/.
2014d Is authority constructed and contextual? *Sense & Reference*, July 22, 2014. Retrieved August 26, 2019 from https://senseandreference.wordpress.com/2014/07/22/is-authority-constructed-and-contextual/.
2014e Is format a process? *Sense & Reference*, July 25, 2014. Retrieved August 26, 2019 from https://senseandreference.wordpress.com/2014/07/25/is-format-a-process/.
2014f Is searching exploration? *Sense & Reference*, July 29, 2014. Retrieved August 26, 2019 from https://senseandreference.wordpress.com/2014/07/29/is-searching-exploration/.
2014g Does information have value? *Sense & Reference*, August 5, 2014. Retrieved August 26, 2019 from https://senseandreference.wordpress.com/2014/08/05/does-information-have-value/.
2016a Revisiting the Framework: Is authority constructed and contextual? *Sense & Reference*, July 19, 2016. Retrieved August 26, 2019 from https://senseandreference.wordpress.com/2016/07/19/revisiting-the-framework-is-authority-constructed-and-contextual/.
2016b Revisiting the Framework: Is information creation a process? *Sense & Reference*, July 22, 2016. Retrieved August 26, 2019 from https://senseandreference.wordpress.com/2016/07/22/revisiting-the-framework-is-information-creation-a-process/.
2016c Revisiting the Framework: Does information have value? *Sense & Reference*, July 29, 2016. Retrieved August 26, 2019 from https://senseandreference.wordpress.com/2016/07/29/revisiting-the-framework-does-information-have-value/.

2016d Revisiting the Framework: Is research inquiry? *Sense & Reference*, August 9, 2016. Retrieved August 26, 2019 from https://senseandreference.wordpress.com/2016/08/09/revisiting-the-framework-is-research-inquiry/.

2016e Revisiting the Framework: Is scholarship a conversation? *Sense & Reference*, August 12, 2016. Retrieved August 26, 2019 from https://senseandreference.wordpress.com/2016/08/12/revisiting-the-framework-is-scholarship-a-conversation/.

2016f Revisiting the Framework: Is searching strategic explanation? *Sense & Reference*, August 17, 2016. Retrieved August 26, 2019 from https://senseandreference.wordpress.com/2016/08/17/revisiting-the-framework-is-searching-strategic-explanation/.

Winchester, S.

1998 *The Professor and the Madman: A tale of murder, insanity, and the making of the Oxford English Dictionary.* HarperCollins, New York.

Dissecting Digital Divides in Teaching

William Caraher

Introduction

My paper considers the impact of so-called digital divides in digital approaches to teaching about the Ancient and Medieval worlds. My experience mostly derives from teaching a large (150+ student) introductory-level "Western Civilization" class at the University of North Dakota. UND is a mid-sized, "High Research Activity" university (according to our Carnegie classification) that draws heavily from the Northern Plains. I teach in small history department of 10 faculty with relatively strong commitment to undergraduate teaching and a small and withering graduate program. This paper explores how various "digital divides" have shaped my own teaching strategies in an introductory level history course and how working to bridge these divides on a practical level nudged me to think more critically about how digital tools produce students, teachers, and communicate the expectations of the modern world.

The Digital Divide

There's been a good bit of scholarship on the digital divide in secondary and higher education. The idea of the digital divide, in its most basic form, suggests that a significant divide exists between those who use and have access to digital technologies and those who do not.[1] This divide is usually mapped along social, economic, and regional lines. Rural states, like North Dakota, tend to fall on one side of the digital divide especially where access to broadband internet is concerned and in terms of how frequently secondary and higher school

---

[1] McConnaughey et al. 1995.

students use computer for homework (Halvorson 2018). In fact, 153 of the 176 school districts in North Dakota are categorized as either "rural, distant" or "rural, remote" by the National Center of Education Statistics, marking them as having the lowest access to broadband internet by a substantial margin, Just across state lines, districts in the northern tier of Minnesota are similarly defined.[2] The impact of the digital divide on rural communities in North Dakota is sufficiently acute that the Chancellor of the North Dakota University System recently floated the idea of a "Cyber-Grant" university initiative that would tax tech companies to help support the development of digital infrastructure in states that lag behind.[3]

Although not all UND students come from rural areas and most, in fact, don't come from North Dakota,[4] my experience is that they are generally less technologically savvy and comfortable in digital environments than their more affluent and more suburban counterparts elsewhere in the U.S.

While the data suggests that students from rural areas are no less likely to own digital devices than their suburban or urban counterparts,[5] I continue to be struck by the significant number of students for whom technology is not a constant companion. Many of my students do not bring their laptops to class regularly, for example. In a recent field project that involved using mobile phones to take video, a number of students had such outdated phones that they could not accommodate more than short video clips; one student had a flip phone; another student took videos but was never able to send them to our archive. While it was easy enough to negotiate the different access to technology, it also was clear that the digital divide in terms of hardware remains firmly in place. A recently updated smart classroom with a series of small group work stations relies on students to use their own laptops to access the large, shared monitor. This seems like an optimistic implementation of technology.

[2] NCES 2018.
[3] Hagerott 2018.
[4] UND Student Profile 2018-2019.
[5] Croft and Moore 2019.

Access to the right hardware, however, is only part of the digital divide. Over the last decade of teaching, it has become clear to me that something as simple as a broken hyperlink or a pdf document oriented the wrong way, can represent a barrier to accessing information. A significant group of students lack the standard tool kit of web "work arounds" that range from savvy web searches to negotiating the standard elements of user interfaces across multiple applications. Such simply life hacks as using a mobile device as a quick and dirty scanner or looking for an article on Academia.edu or institutional repositories remains on the fringes of their practice (even when such approaches are modeled in class). Finding ways to access pirated copies of publications or books to accelerate research, whatever the ethical and legal risks of such practice, is simply beyond what we can expect.

In my larger experience across campus at UND, it is pretty apparent that even relatively simply digital interfaces—like editable Wikis or shared documents in Google or Microsoft 365—caused myriad small-scale obstacles that frustrated students and complicated group work.

Prosumer and Consumers

Access to hardware and familiarity with software (and these often go hand-in-hand) sketches one level of the digital divide; these also contribute to the existence of what some scholars have called the second-level digital divide.[6]

The second level divide maps the difference between individuals who are consumers of digital material on the web and those who are so-called prosumers of digital and web-based content.[7] Prosumers both consume and produce products, content, and media on the web and so-called "prosumption" is the backbone to the participatory web, Web 2.0, and, in some ways, anticipates the semantic web (or Web 3.0) outlined by Tim Berners-Lee and embraced by so many archaeologists.[8] The lag in access to broadband may well have had a much

---

[6] Hargittai 2002.
[7] Toffler 1980; Tapscott and Williams 2006.
[8] For the term semantic web see Tim Berners-Lee 2001; for its use in an archaeo-

greater impact on students' ability to see the digital world as a space of shared media, data, and experiences. While my students do have social media accounts, they tend to be skeptical of blogging, consume YouTube and podcasts more than produce in these media, and are particularly hostile toward Wikipedia.

I contend that this second level divide is far more problematic than the first level divide for implementing digital approaches to teaching and, as a result, I have dedicated more time to cultivating prosumer culture among my students and demonstrating how digital tools facilitate certain kinds of collective knowledge making. My approach to bridging the second-level digital divide, however, is intentionally naive in order to mask my deeper ambivalence about it. On the one hand, I continue to have a certain amount of faith that the last unfettered wilds of the internet hold out a glimmer of hope for a society and at times feel inspired by works like Michael Serres refiguring of *Thumbelina*.[9]

On the other hand, I worry that at the current moment, the digital world is contributing to a society that is far more likely to be shackled, monitored, and manipulated by technology than liberated by it. I want my students to understand the power of Wikipedia, for example, and the ecosystem that has produced the growing number of open educational resources and open-source software, and the potential, if not unproblematic character, of maker culture.[10] Moreover, I want them to be prepared to contribute to it.

At the same time, I do recognize that most aspects of prosumer culture have been coopted by the usual suspects of capitalism,[11] sexism,[12] racism,[13] and technological solutionism.[14] By producing new knowledge, creative works, and tools, we also produce profits for

---

logical context see Kansa 2014.
[9] Serres 2015.
[10] Chachra 2015.
[11] Fuchs 2013.
[12] Glott and Ghosh 2010; Losse 2014; Leonard and Bond 2019.
[13] O'Neil 2009; Montez 2017.
[14] Morozov 2013.

transnational corporations who are as comfortable limiting access to our own work as they are preventing us from subverting their spirit of profit. As the kids say: "the revolution will now be monetized."[15]

Other Digital Divides

The digital divide and the consumer/prosumer divide are similar to older, more persistent, and equally porous divides that structure how we learn and think. In my discipline of history, students obsess over and are baffled by the distinction between primary and secondary sources. For students of the ancient Mediterranean, this consternation is particularly understandable and useful for unpacking the relative uselessness of this distinction among practicing historians. A source is a source and only primary or secondary in relation to how it is used, or to paraphrase E.H. Carr evidence is only evidence when its evidence for something.[16]

Practicing archaeologists sometimes find ourselves in the same bind, of course. The divide between data and interpretation, for example, coincides with the primary and secondary source divide among historians. The persistence of terms like "raw data" reveals an understanding of archaeological knowledge-making that divides data from interpretation.[17] It seems to me that digital data makes this divide all the more convenient in part because the data itself appears so distinct from interpretative texts, and partly because "digging down" into the data (or data mining) represents a useful play on the modernist assumption that excavation (literally or metaphorically) provides access to a view of the past less encumbered by present interpretation. While we may intellectually understand this divide as naive as generations of archaeologists who celebrate reflexivity and methodology have taught us, we nevertheless tend to lean on the distinction between data and interpretation to frame our conversations. Endless references to archaeological data populate academic conferences, publications, and, I suspect, our teaching. For students who

---

[15] Zimmerman 2017.
[16] Carr 1961.
[17] Gitelman and Jackson 2013.

continue to want to see facts as the antidote to fake news, the transparent use of data appears to be a compelling ontological tonic for their epistemological anxiety.

To my mind, this digital divide is every bit a pernicious as the other digital divides that shape contemporary culture. In fact, it might be more dangerous in the era of Big Data than the other digital divides because it tends to see data as holding a particular kind of fundamental and inescapable authority in how it describes the world.

A Critique of Prosumption

All of this brings me to my Introduction to Western Civilization class at the University of North Dakota, which I have taught for the last five years in a Scale-Up style classroom. The idea behind Scale-Up classrooms originated at NC State and the term "Scale-UP" was an acronym for "Student-Centered Activities for Large Enrollment Undergraduate Physics."[18] Today, folks talk about "Student-Centered Active Learning Environment with Upside-down Pedagogies," but the general idea remains that these classrooms are designed to accommodate large classes with flipped pedagogies.

My Western Civilization class generally enrolled 150-180 students and the room was set up for them to sit around round, 9-person tables. Each table had three laptops connected to a monitor and also came with a whiteboard and a microphone for the students to play with when bored. A central teaching station allowed me to observe most of the groups and to project content from the tables onto four large projection screens in the corners of the room. The goal of this class was for students to become better at making sustained arguments about the past and to do this at scale.[19]

The design of the room encouraged students to work together and at least in theory sought to mitigate the hardware aspects of the digital divide by ensuring that at least three students had access to a laptop. In the most common implementations of this design, a

---

[18] Gafney et al. 2008.
[19] Some of the ideas that I presented here appeared originally in an unpublished paper by Caraher and Stanley 2014.

student or students worked as the scribe for the table on a provided laptop or students worked in smaller groups, three to a laptop, sometimes installed with software appropriate the assignment or the discipline. While I did not formally leverage the practical aspects of three-laptop design, it did work to level uneven access to technology among my students.

The class sought to bridge the "second-level digital divide" by encouraging students to work critically as prosumers of educational content. In practice, this involved having the students write a Western Civilization textbook with each table working on a series of chapters over the course of the semester that we bring together at the end of the class as a completed book. This task encouraged students to recognize the value of their own voice, critical abilities, and maybe even responsibility to produce their own historical narratives and analysis. It also subverts some of the economic and political power of textbook publishers; although, I do ask them to buy a used copy of an older version of a textbook as a model.

Finally, the students start with a more or less a blank document. I do not provide an approved list of primary or secondary sources or even offer much in the way of a critical guide to navigating the internet. Most students get that journal articles are better than random webpages (of uncertain authorship and content), that Wikipedia is a good place to glean chronology, geography, and additional sources, and that historical arguments are only as good as the sources that they identify to build their arguments. If they cannot find good evidence for an argument, then no amount of rhetorical savvy is likely to make it compelling.

I use this approach as a way to de-emphasize the idea that there is a body of data "out there" ready for consumption, analysis, and interpretation. Instead, it encourages the students to see the body of useful evidence and data as the product of their research questions and priorities. The "raw material" of history is not something that is "mined" for knowledge, but something that is created as evidence for arguments about the past.

In an era where relational data is literally being treated and traded as a commodity, it is hardly surprising that we envision knowledge making as a kind of extractive industry rather than, say, performative or generative (and, here, I'm inspired by a paper that my colleague Sheila Liming gave a few years back on the metaphor of data and text mining.[20])

I guess that I should admit that this class is chaotic in every way. The groups struggle mightily with finding sources, producing specific evidence, and making arguments. We produce outlines, write drafts, have peer reviews, and revise, all the while careening closer the goal of a careful argument. I like to imagine that the uneven results, the frustration, and the chaos reproduces the struggle that most of us have in making sense of the digital world and constructing arguments. By trying to break down the divisions that my students imagine between data and argument, evidence and assertion, or even fact and fake news, we accept more readily the messy and complicated state of knowledge in the real world.

Conclusion

This paper does not have some kind of brilliant and inspiring TED talk style conclusion. In fact, I would be remiss if I did not point out that the prosumer culture in the Scale-Up classroom has its own economic, political, and social baggage. My class prepares students to live in a world populated by Uber drivers, to repurpose apartments as Air BnBs, and to celebrate so-called "maker culture" that is far from being radical. In fact, collaborative styles of learning may simply reorder many sexist, classist, and racists features of 20th century industrial capitalism. After all, as Arum and Roksa (2011) pointed out almost a decade ago, collaborative learning models tend to privilege more affluent students from more educated families.

For example, I recognize that some of the control that I visibly cede to the students, is an illusion that parallels many of the illusory aspects of freedom and control central to our digital culture. The

---

[20] Liming 2016.

digital world has made observation central to how we monetize time in late capitalism (page views, active time, engagement time, et c.).[21] This same approach is baked into most learning management systems which allow us to track student activities when they visit our class site. Moreover, there is a kind of panopticism inherent in the design of the Scale-Up room. While the students face one another, I stand in a position that allows me to observe the dynamic in groups and across that classroom. The role of teacher as observer is central to understanding what some have called "invisible learning"[22] or "intermediate processes" central to the acquisition of higher-level thinking skills. Our ability to observe both the analog work of students in groups as well as their digital work (through the backend of our learning management system) contributes to a 21st century version of the kind of surveillance society that Foucault identified as characteristic feature of the modern world.

While it might sound naive to assume that as thoroughly a modern discipline as education, could avoid inculcating students with the expectations of the market, I do worry that our own use of digital tools and environments do little to prepare students to resist these pressures. On the other hand, perhaps an encounter with the digital world based around dissection and breaking down these digital divides at least offers a tool kit for students to expect there to be limits to practices and to engagement in the digital world. This, of course, does nothing to undermine an ironic view of the modern world where strategies of dissimulation and occlusion obscure the real function of power and the making of meaning. At the same time, for as long as there has been formal education, students have found ways to resist the expectations of the classroom, our institutions, and our pedagogy. We can hope that this resistance is more than just pushing back against authority or against the discomfort of learning, but an informed resistance to the system itself.

---

[21] Fuchs 2013.
[22] Cobo and Moravec 2011.

Works Cited

Arum, Richard and Josipa Roksa
    2011 *Academically Adrift: Limited Learning on College Campuses*. University of Chicago Press, Chicago.

Berners-Lee, Timothy
    2001 *Weaving the Web: The Original Design and Ultimative Destiny of the World Wide Web by its Inventor*. HarperBusiness, New York.

Caraher, William and C. Stanley
    2014 Teaching History in a Scale-Up (Student-Centered, Active Learning Environment for University Programs) Classroom: Some Reflections on Method and Meaning. Unpublished manuscript. http://dx.doi.org/10.17613/M6D21RJ25

Carr, E.H.
    1961 *What is History?* Vintage Books, New York.

Chachra, Debbie
    2015 Why I'm not a Maker. *Atlantic* Online. 23 January https://www.theatlantic.com/technology/archive/2015/01/why-i-am-not-a-maker/384767/

Croft, Michael and Raeal Moore
    2019 *Rural Students: Technology, Coursework, and Extracurricular Activities. Insights in Education and Work*. February. ACT Research and Center for Equity in Learning. Iowa City, Iowa. https://equityinlearning.act.org/wp-content/themes/voltron/img/tech-briefs/rural-students.pdf

Gaffney, Jon D.H., Evan Richards, Mary Bridget Kustusch, Lin Ding, and Robert J. Beichner,
    2008 Scaling Up Educational Reforms. *Journal of College Science Teaching* 37(5): 18-23

Gitelman, Lisa and Virginia Jackson
    2013 Introduction. In *"Raw Data" is an Oxymoron*. Edited by Lisa Gitelman, pp. 1-14. MIT Press, Cambridge, Massachusetts.

Ruediger Glott and Rishab Ghosh
    2010 *Analysis of Wikipedia Survey Data: Topic: Age and Gender Differences*. Collaborative Creativity Group, Wikimedia Foundation, United Nations University MERIT, Maastricht Univer-

sity, March. https://web.archive.org/web/20100414165445/
http://wikipediasurvey.org/docs/Wikipedia_Overview_
15March2010-FINAL.pdf.

Hagerott, Mark
2018 Silicon Valley Must Help Rural America. Here's How. *Chronicle of Higher Education*. September 23. https://www.chronicle.com/article/Silicon-Valley-Must-Help-/244573

Halvorson, Rusty
2018 Forum in Bismarck Examines North Dakota's Digital Divide. American Ag Radio Network. January 12. https://americanagnetwork.com/2018/01/forum-bismarck-examines-north-dakotas-digital-divide/

Hargittai, Eszter
2002 The Second Level Digital Divide. *First Monday* 7(4). https://firstmonday.org/ojs/index.php/fm/article/view/942

Institute of Educational Sciences. National Center for Educational Statistics.
2018 *Rural Education in America*. https://nces.ed.gov/surveys/ruraled/

Kansa, Eric
2014 Open Context and Linked Data. *ISAW Papers* 7(10). http://dlib.nyu.edu/awdl/isaw/isaw-papers/7/kansa/

Liming, Sheila
2016 Fracking the Canon: Spatial Metaphors and the Stakes of Invasive Critique. Paper presented at the 2016 Modern Language Association Convention. Austin, Texas.

Losse, Kate.
2014 Sex and the Startup: Men, Women, and Work. *Model View Culture*. March 17. https://modelviewculture.com/pieces/sex-and-the-startup-men-women-and-work

Leonard, Victoria and Sarah Bond
2019 Advancing Feminism Online: Online Tools, Visibility, and Women in Classics. *Studies in Late Antiquity* 3(1): 4-16. DOI: 10.1525/sla.2019.3.1.4

McConnaughey, Jim, Cynthia Ann Nila, and Tim Sloan
    1995 *Falling Through the Net: A Survey of the "Have Nots" in Rural and Urban America*. National Telecommunications and Information Administration: Washington, D.C. https://www.ntia.doc.gov/ntiahome/fallingthru.html

Montez, Noe
    2017 Decolonizing Wikipedia through Advocacy and Activism: The Latina/o Theatre Wikiturgy Project. *Theatre Topics* 21(1): E-1-E-9.

O'Neil, Mathieu
    2009 *Cyberchiefs: Autonomy and Authority in Online Tribes*. Pluto Press. New York.

University of North Dakota
    2019 UND Student Profile. https://www1.und.edu/research/institutional-research/student-profile/

Zimmerman, Amy
    2017 The Revolution Will Be Monetized: Kendall Jenner, Ivanka Trump and Taylor Swift's Faux-Feminism. *The Daily Beast*. April 6. https://www.thedailybeast.com/the-revolution-will-be-monetized-kendall-jenner-ivanka-trump-and-taylor-swifts-faux-feminism

Autodidacts and the "Promise" of Digital Classics

Patrick J. Burns

Introduction

The *Digital Approaches to Teaching the Ancient Mediterranean* conference at the Institute for the Study of the Ancient World addressed the pedagogical concerns of an admirable array of ancient-world topics and, at least with respect to higher education, pitched these concerns to a broad range of institutions, including so-called R1 universities, large state institutions, large private institutions, small liberal arts colleges, even humanities think tanks (like the event's host institution). Despite this broad coverage, one audience that was absent from the official program is the admittedly hard to classify and hard to quantify group of people who develop an interest in the ancient Mediterranean and seek to educate themselves on its languages, culture, history, and related fields, that is ancient-world autodidacts. As the conference's proceedings demonstrate, digital approaches have had a significant impact on college-level teaching of ancient-world topics, but for autodidacts the impact is perhaps even greater, even more transformative. Paradigm shift may not be too strong a description.

In this brief response to the *DATAM* conference, I consider the audience for Digital Classics research outside of the academy—in fact, outside of formal education altogether—namely, independent learners who are able to use our publications, platforms, tools, and datasets to teach themselves about the ancient world.[1] I argue that

---

[1] Much of what I have to say here is sympathetic with arguments found throughout Gabriel Bodard and Matteo Romanello's collected volume, *Digital Classics Outside the Echo-Chamber* (Bodard and Romanello 2016), especially those parts where "arguments around public engagement, reception, crowdsourcing and citizen science" (3) are addressed; particular chapters of interest include Mahony 2016, Rydberg-Cox 2016, and Almas and Beaulieu 2016.

the dominant ethos of open-source development and open-source distribution in Digital Classics demonstrates "promise" to this audience in two important ways: 1. it represents our fulfillment of a contract that our research output should be a contribution to knowledge in general (as opposed to a contribution for a select academic audience); and, 2. it activates the learning potential of an audience who for a variety of reasons will not become our students in a formal educational context.[2]

I will restrict my comments largely to digital approaches to "Classics," by which I mean the (perhaps overly narrow) study of Ancient Greek and Latin language and literature, because that is the area of "teaching the Ancient Mediterranean" that I know best.[3] That said, the larger point stands for ancient-world study in general and I invite my colleagues working in archaeology, numismatics, papyrology, epigraphy, and so on, as well as in languages beyond Greek and Latin, to reflect on who their audience of autodidacts may be and how their scholarly output may support in a substantially similar way these students outside the academy proper.

A personal anecdote to begin—I was once an ancient-world autodidact. My career in Classics began with teaching myself Latin in my late 20s from a cobbled-together collection of print textbooks, grammars, lexica, readers, and so on. At some point—as I suspect is the case for nearly all Classics students in this century—I found the Perseus Digital Library.[4] It was a profound and confusing epiphany.

---

[2] For a related discussion of the "promise" of digital resources, see Smith and Casserly 2006 ("the promise of open educational resources"). Thomas 2015 (on "the promise of the digital humanities") discusses the institutional requirements necessary to allow fully digital humanists to "take advantage of the networks, spaces, and audiences online to create and refine new forms of...scholarship" (534).

[3] On broadening the definition of "Classics," see, for example, Quinn 2018; Levine 1992; and the mission statement of *DATAM*'s host institution, the Institute for the Study of the Ancient World, available at http://isaw.nyu.edu/about.

[4] See Crane 1998: "Even now, as our modest digital library on ancient Greek culture finds its way into homes, schools and offices where traditional scholarly publications have not reached, we can see by the patterns of use and the mail that we receive the stirrings of a vast audience." On classical language learning in this context specifically, see Rydberg-Cox 2016: 79.

Why were all of these texts here? Why was I able to click a word and see its definition? Why was this all free? (I had yet to consider the cost of "free."[5]) Who was doing this work? I spent so many hours on Perseus in these formative years of my training that it is not an exaggeration to say that I felt at times like an unofficial Classics major at Tufts.

What I did not realize then was that what was on offer at Perseus was part of a movement taking hold in the 1990s and coming into its own in the following decade, namely open-access publication, or freely available, internet-distributed content.[6] Open-access content is of course not restricted to educational content, although institutions of higher education were in a particularly good position to produce materials at an early stage.[7] Universities, for example, were already producers of knowledge with access to relatively high-speed connections, sufficiently ample storage, and often their own dedicated servers. In addition, as knowledge creators and teachers already in their vocational disposition, they also had an available audience of a similarly provisioned research community across institutions as well as enrolled students. Materials produced under these conditions for these audiences are the resources which would define the Open Educational Resources (OER) movement in the early 2000s.[8] Soon every discipline would have a Perseus (or more accurately, many Perseuses) providing academic content online.

---

[5] See Kamenetz 2010: 104–106 on the costs of open resources and the role of institutional support in mitigating these costs.

[6] A fuller definition of "open access" provided by the Budapest Open Access Initiative (Chan et al. 2002) is still serviceable: "Free availability on the public internet, permitting any users to read, download, copy, distribute, print, search, or link to the full texts of these articles, crawl them for indexing, pass them as data to software, or use them for any other lawful purpose, without financial, legal, or technical barriers other than those inseparable from gaining access to the internet itself. The only constraint on reproduction and distribution, and the only role for copyright in this domain, should be to give authors control over the integrity of their work and the right to be properly acknowledged and cited." For a review of other definitions, see Bailey Jr. 2007.

[7] On the role of open access in knowledge production and the "intellectual commons," see Suber 2006.

[8] For an overview of open educational resources, see Wiley, Bliss, and McEwen

A consequence of the widespread availability of open-access materials is that they found readily an audience outside of academia. These resources are the foundations upon which "Do-It-Yourself University," as Anya Kamenetz would describe it, rests: a "complete educational remix," the "expansion of education beyond classroom walls," and the possibility of "free, open-source, vocational, experiential, and self-directed learning."[9] Whether in the form of highly organized, institutionally-backed efforts, like MIT OpenCourseWare and the wide array of massive open online courses (MOOCs) that came into their faddish own around 2012, or smaller, distributed efforts like a professor making a course syllabus available on their academic website, opportunities to learn online using open content had become and would remain ubiquitous.[10]

Classics has participated in this open education movement for decades now. The Perseus Project stands out not only as an early player in an internet-based Classics, but because of editor-in-chief Gregory Crane's embrace of open-source development for the Perseus software and embrace for open publication standards for its content, not to mention the voluminous writings by Crane and his collaborators defending this position and advocating for its democratizing, access-expanding potential. Furthermore, Perseus was not alone here. A look at the table of contents for the three-year run (1998-2000)

2014. The idea of "promise" has been built into the OER vision from the beginning; see Tuomi 2006: 3: "Assume a world where teachers and learners have free access to high-quality educational resources, independent of their location....In the next several years, it will become possible in a scale that will radically change the ways in which we learn and create knowledge."

[9] Kamenetz 2010: x.

[10] On OpenCourseWare, see Abelson 2008. On the explosion of popularity of MOOCs, see Pappano 2012. While it is more difficult to pinpoint the direct effects of the "smaller, distributed efforts," it is worth reminding ourselves of just how novel these were just twenty years back; see, for example, Small 1998, a review of personal web pages in the short-lived *Bryn Mawr Electronic Resources Review*. Small's review incidentally refers to the "wonderful omnium gatherum" of archaeological resources on the internet by *DATAM* organizer, Sebastian Heath, evidence of a two-decade commitment to the intersection of digital resources and ancient-world studies.

of the *Bryn Mawr Electronic Resources Review* provides a convenient snapshot of early efforts in the field to find an audience outside of the academy, including projects that still offer substantial resources for an autodidact audience like *VRoma* and *LacusCurtius*, to name just two. The audience for these platforms comes across, for example, in William Hutton's review of the *Diotima* project, where he writes that the site's content on women and gender in antiquity is "potentially of use to anyone with even glancing interest in the ancient world."[11]

Here lies the first "promise" with respect to autodidactism mentioned above. At the core of Digital Classics's commitment to open resources in its formative stages was an obligation to make the discipline available to as wide an audience as possible. To put this another way, although the research and teaching output may have originally been aimed at an academic audience, a superseding responsibility emerged and continues to be a prevalent mindset among Digital Classicists that it is incumbent on us to provide materials to the "wider community" of learners.[12] As writer and English instructor Kim Wells once wrote on her "fan site [for] canonically excluded women writers," *Domestic Goddess*: "I think it is our duty as teachers not to ignore the possibilities of making research easily available on the Internet. If educators do not provide the information, who will?"[13]

---

[11] Hutton 1999.

[12] Blackwell et al. 2006: "Immense digital libraries based on open access and aimed at massive audiences put scholars under an obligation to avoid a new access divide opening up between ourselves and the wider community that we serve." It should be added that there is also often motivation based on financial reciprocity here: we contribute to the public because the public has invested in us. So, Romanello and Bodard 2016: 8: "Since academic research is largely funded by public money, it is arguably incumbent upon us to find ways to engage the public with our findings."

[13] Wells 2000, as quoted in Earhart 2015: 73. There is a worthwhile, if sadly ironic, lesson with respect to Wells's quote about "research [made] easily available on the Internet." Wells's *Domestic Goddess* is no longer available at the URL www.womenwriters.net; this URL now points to the website for an essay writing service. Thankfully, because of the efforts of the Internet Archive and its Wayback Machine, Wells's contributions have been preserved. Nevertheless, this example does point to fragility in the system and argues again for the public benefit and contribution to knowledge that institutional investment in open resources provides.

The second "promise" I wish to discuss here is the promise latent in an audience interested in various aspects of the Classical world, but who for a variety of reasons will not become our students in a formal educational context.[14] It is unnecessary to rehearse here the multiple and various barriers to higher education.[15] Suffice it to say that these barriers can be compounded within a humanities discipline such as Classics. Students faced with limited time and an imposing dollar-per-credit ratio may feel pressure to take more "useful" courses with "better job prospects" in their formal course of study.[16] No less consequential are the systemic barriers, such as classism, sexism, racism, and ableism, that have long restricted access to Classics as a discipline.[17] The ubiquitous, on-demand, asynchronous, and largely cost-free offerings that are now available, in no small part due to open-source development and open-access publication among Digital Classicists, are by no means a panacea, but they can make significant inroads in expanding the community of potential learners and so, by extension, expanding the Classics community in general.

When I joined Twitter in 2011, I chose the handle @diyclassics as a nod to the widespread do-it-yourself (D.I.Y.) ethos in American hardcore punk focused on making music outside of a traditional corporate model.[18] In my case, I was beginning to think about what Classics could look like outside of a traditional institutional model.[19] For the

---

[14] Smith and Casserly 2006.

[15] For a discussion of barriers to higher education, see, for example, Page and Scott-Clayton 2015.

[16] Schmidt 2018. It should be noted that Schmidt qualifies the "better job prospect" idea in the article, writing: "Students aren't fleeing degrees with poor job prospects. They're fleeing humanities and related fields specifically because they *think* they have poor job prospects" (emphasis in the original).

[17] For a starting point on systemic barriers within Classics, see Adler 2017, as well as important contributions from Bracey 2017, Chae 2018, "Sankarshana" 2019, Erny, Nakassis, and Steinke 2017, and Sharples 2019, to name just a few recent examples from the online journal *Eidolon*, a leading voice during an active moment of self-reflection and critique within the discipline concerning these barriers. See also Morley 2018: 37–38.

[18] On the core values of D.I.Y., including the role of technology and the internet in reshaping these values, see Moran 2010: 62–63.

[19] For a similar way of thinking about an adjacent field, see the "punk archaeolo-

musicians, it was a matter of "taking control."[20] My vision for a D.I.Y. Classics coalesced around the field's digital output and had a similar aim. Whether it was complete ancient Greek and Latin texts through the *The Latin Library*, *LacusCurtius*, or the *Perseus Project*, dictionaries and other reference materials through *Perseus*, *Logeion*, or the *Suda Online*, even cutting-edge scholarly communication through sites like *Bryn Mawr Classical Reviews*, *Greek, Roman, and Byzantine Studies* or *Classics@*, the building blocks for an autodidactic model for Classics were falling into place.[21] Someone curious about the discipline did not have to rely wholly on being taught but rather was empowered to learn on their own terms. It is true that I continued to pursue formal graduate training in Classics, but I did so in a way that fostered the "promises" described above. I am now a practicing Digital Classicist, building open-source tools and distributing open-access materials, so that the current (and future) generation of ancient-world autodidacts can pursue their studies.

Simply stated, we have an opportunity to make an enormous contribution to our discipline by acknowledging our autodidact audience and making materials available to them. This is important pedagogical work. It is also important outreach work. The digital resources—and specifically the open digital resources—presented and discussed during *DATAM* are contributions to the field which foster curiosity and engagement in the objects of our study and increase the number of people who can "contribute to a discussion of what Classics is and what it might be."[22]

---

gy" essays in Caraher, Kourelis, and Reinhard 2014.

[20] See Azerrad 2001: 6: "'Punk was about more than just starting a band,' former Minutemen bassist Mike Watt once said, 'it was about starting a label, it was about touring, it was about taking control.'" McManus and Rubino 2003: 601 mention "control over learning" as an advantage of Digital Classics pedagogy.

[21] *GRBS* became an open-access journal in 2010 on "the principle that making research freely available to the public supports a greater global exchange of knowledge"; see https://grbs.library.duke.edu/about/editorialPolicies.

[22] This quote is taken from the "Outreach" page on the website of the Society for Classical Studies (https://classicalstudies.org/outreach/home). Another key sentence: "At this exciting moment, the multitude of new technologies and modes of communication can make it easier than ever before to connect with the great

Thankfully, this essay is *not* a call to action. I think that it would be fair to say that open-source development and open-access distribution are the dominant practices of Digital Classics, a vanguard led by Perseus for decades now and adopted as received wisdom by much of the community since. So, not a call to action, but rather a reminder that, because of our embrace of open resources and our commitment to making them widely available, Digital Classicists have students who we never see, but whose studies are enriched by our work. Reciprocally, our field is enriched by their interest and participation, and this is a phenomenon worth noting in a discussion of digital approaches to teaching the Ancient Mediterranean.

---

achievements of the past and their meanings for us now." This gesture toward outreach relates to discussions of public scholarship in general. These discussions are deep and wide-ranging; the just released *The Oxford Handbook of Methods for Public Scholarship* edited by Patricia Leavy (Leavy 2019) looks promising in providing a systematic overview. Lastly, there is good Classics outreach work being done right now in the United Kingdom that deserves mention in this context; see Holmes-Henderson, Hunt, and Musié 2018, and especially the chapter by James Robson and Emma-Jayne Graham (Robson and Graham 2018) which covers the role of open-access materials at the Open University.

Works Cited

Abelson, H.
2008 The Creation of OpenCourseWare at MIT. *Journal of Science Education and Technology* 17(2): 164–174.

Adler, E.
2017 On Classism in Classics. *Eidolon* Nov. 13, 2017. https://eidolon.pub/on-classism-in-classics-157c5f680c4a

Almas, B., and M.C. Beaulieu
2016 The Perseids Platform: Scholarship for all! In *Digital Classics Outside the Echo-Chamber: Teaching, Knowledge Exchange and Public Engagement* edited by G. Bodard and M. Romanello, pp. 171-186. Ubiquity Pres, London.

Azerrad, M.
2001 *Our Band Could Be Your Life*. Little, Brown, and Co., Boston.

Bailey Jr., C. W.
2007 What Is Open Access? *Digital Scholarship* Feb. 7, 2007. http://digital-scholarship.org/cwb/WhatIsOA.htm

Blackwell, C., G. Crane, H. Dik, C. Roueché, J. Rydberg-Cox, R Scaife, N. Smith, and H. Cross
2006 *Classics in the Million Book Library*. http://www.stoa.org/million/chicagostatement.html

Bodard, G., and M. Romanello, M. (editors)
2016 *Digital Classics Outside the Echo-Chamber: Teaching, Knowledge Exchange and Public Engagement*. London.

Bracey, J.
2017 Why Students of Color Don't Take Latin. *Medium* Oct. 12, 2017. https://eidolon.pub/why-students-of-color-dont-take-latin-4ddee3144934

Caraher, W. R., K. Kourelis, and A. Reinhard (editors)
2014 *Punk Archaeology*. The Digital Press at the University of North Dakota, Grand Forks, N.D.

Chae, Y. I.
2018 White People Explain Classics to Us. *Eidolon* Feb. 5, 2018. https://eidolon.pub/white-people-explain-classics-to-us-50ecaef5511

Chan, L., D. Cuplinskas, M. Eisen, F. Friend, Y. Genova, J.-C. Guédon, M. Hagemann, M., et al.
  2002 Read the Budapest Open Access Initiative. *Budapest Open Access Initiative* Feb. 14, 2002. https://www.budapestopenaccessinitiative.org/read

Crane, G.
  1998 The Perseus Project and Beyond: How Building a Digital Library Challenges the Humanities and Technology. *D-Lib Magazine* 4(1). http://dlib.org/dlib/january98/01crane.html

Earhart, A. E.
  2015 *Traces of the Old, Uses of the New: The Emergence of Digital Literary Studies*. University of Michigan Press, Ann Arbor, MI.

Erny, G. D. Nakassis, and S. Steinke
  2017 Cleaning Up the Field (S. Scullin interviewer). *Eidolon* Dec. 26, 2017. https://eidolon.pub/cleaning-up-the-field-f7c4c15a2f08

Holmes-Henderson, A., S. Hunt, and M. Musié (editors)
  2018 *Forward with Classics: Classical Languages in Schools and Communities*. Bloomsbury Academic, London.

Hutton, W.
  1999 Review: *Diotima*: Materials for the study of women and gender in the ancient world, by S. Bonefas and R. Scaife, *Bryn Mawr Electronic Resources Review* Jul. 31, 1999. http://bmcr.brynmawr.edu/bmerr/1999/HuttoDiotiSep.html

Kamenetz, A.
  2010 *DIY U: Edupunks, Edupreneurs, and the Coming Transformation of Higher Education*. Chelsea Green Publishing, White River Junction, VT.

Leavy, P.
  2019 *The Oxford Handbook of Methods for Public Scholarship*, Oxford University Press, Oxford.

Levine, M. M.
  1992 Multiculturalism and the Classics. *Arethusa* 25(1): 215–220.

Mahony, S.
2016 Open Education and Open Educational Resources for the Teaching of Classics in the UK. In *Digital Classics Outside the Echo-Chamber: Teaching, Knowledge Exchange and Public Engagement* edited by G. Bodard and M. Romanello, pp. 33-50. Ubiquity Pres, London.

McManus, B. F., and C.A. Rubino
2003 Classics and Internet technology. *The American Journal of Philology* 124(4): 601–608.

Moran, I. P.
2010 Punk: the do-it-yourself subculture. *Social Sciences Journal* 10(1): 58–65.

Morley, N.
2018 *Classics: Why It Matters*. Polity Press, Cambridge.

Page, L. C., and J. Scott-Clayton
2015 Improving College Access in the United States: Barriers and Policy. *NBER Working Papers Series* Working Paper 21781. http://www.nber.org/papers/w21781

Pappano, L.
2012 Year of the MOOC. *The New York Times*, sec. Education Life, Nov. 2, 2012. https://www.nytimes.com/2012/11/04/education/edlife/massive-open-online-courses-are-multiplying-at-a-rapid-pace.html

Quinn, J. Q.
2018 Time to move on. *Times Literary Supplement* Sep. 18, 2018. https://www.the-tls.co.uk/articles/public/time-to-move-on/

Robson, J., and E.-J. Graham
2018 Classics Online at the Open University: Teaching and Learning with Interactive Resources. In *Forward with Classics: Classical Languages in Schools and Communities*, edited by A. Holmes-Henderson, S. Hunt, and M. Musié, pp. 217-230. Bloomsbury Academic, London.

Rydberg-Cox, J.
2016 An Open Tutorial for Beginning Ancient Greek. In *Digital Classics Outside the Echo-Chamber: Teaching, Knowledge Exchange and Public Engagement* edited by G. Bodard and M. Romanello, pp. 69–82. Ubiquity Pres, London. London.

"Sankarshana"
2019 "The Board Is Well-Reminded". *Eidolon* Apr. 4, 2019. https://eidolon.pub/the-board-is-well-reminded-71601252e57d

Schmidt, B.
2018 The Humanities Are in Crisis. *The Atlantic* Aug. 23, 2018. https://www.theatlantic.com/ideas/archive/2018/08/the-humanities-face-a-crisisof-confidence/567565/

Sharples, A.
2019 Disabling Ableism in Classics. *Eidolon* May 20, 2019. https://eidolon.pub/disabling-ableism-in-classics-4fd28d02c628

Small, J. P.
1998 Review of "Personal Web Pages." *Bryn Mawr Electronic Resources Review* Jul. 27, 1998. http://bmcr.brynmawr.edu/bmerr/1998/SmallPersoAug.html

Smith, M. S., and C. M. Casserly
2006 The Promise of Open Educational Resources. *Change: The Magazine of Higher Learning* 38(5): 8–17.

Suber, P.
2006 Creating an intellectual commons through open access. *Understanding Knowledge as a Commons: From Theory to Practice*. https://dash.harvard.edu/handle/1/4552055

Thomas, W. G.
2015 The Promise of the Digital Humanities and the Contested Nature of Digital Scholarship. In *A New Companion to Digital Humanities* edited by Susan Schreibman; Raymond George Siemens; John Unsworth, pp. 524–537. Wiley/Blackwell, Boston.

Tuomi, I.
2006 *Open Educational Resources: What They Are and Why Do They Matter*. http://www.meaningprocessing.com/personalPages/tuomi/articles/OpenEducationalResources_OECDreport.pdf

Wells, K.
 2000 About this site..., *Domestic Goddess* Nov. 9, 2000. https://web.archive.org/web/20001109235300/http://www.womenwriters.net:80/domesticgoddess/about.html

Wiley, D., T. J. Bliss, and M. McEwen
 2014 Open Educational Resources: A Review of the Literature. In *Handbook of Research on Educational Communications and Technology* edited by J. M. Spector, M. D. Merrill, J. Elen, and M. J. Bishop, pp. 781–789. Springer, New York.

Playing the Argonauts: Pedagogical Pathways through Creation and Engagement in a Virtual Sea

Sandra Blakely

Day 3: Night falls around your ship, a dark sky illuminated by multiple constellations above the outline of the Aegean coast. Passing the headland of Tisaia, you watch for the crags of Pelion to guide your way: you are heading northwest to the Tomb of Dolops. Suddenly one crewmember falls ill, then another – the ship's water has gone foul, and you lose five men before you can get to port. Are your funds sufficient to hire their replacements?

Day 7: Your crew suddenly feels uneasy, so you drop anchor, though there are no pirates or storms in sight. After prayers to Poseidon for your good fortune, a group of dolphins jump about the ship, playing for a moment before disappearing. Your crew takes it as a good sign and their spirits are lifted. As they begin to raise anchor, you notice the ship feels a bit faster than before. The waters seem to push you forward: this is both fortunate and suspicious!

These captain's log style summaries offer a glimpse into *Sailing with the Gods*, an online, interactive 3-D game developed to harness the capacity of serious games to generate meaningful data on real world phenomena.[1] The phenomenon is maritime mobility in the Hellenistic

---

[1] The latest iteration of *Sailing with the Gods* is available at https://scholarblogs.emory.edu/samothraciannetworks/the-game/ . This game would be impossible without the time, creativity and energy of every member of the team. Lead Programmer and Simulation Designer, Robert Bryant, University of Pennsylvania; Developer, Kevin Dressel, founder of Shiny Dolphin Games LLC http://www.

world: the setting is a sea humanized by social and civic networks, local myths and ritual traditions, and characterized by the uneven distribution of resources that made connectivity a fundamental Mediterranean strategy (Bresson 2015: 31-41). The pedagogical potential of the project quickly became apparent, even in the early stages of the game's release. Harnessing that potential has engaged the project with critical issues in the world of digital game based learning, and highlighted both the potential and the caveats for the intersection of gaming and education in the ancient world. In this brief paper, I offer an overview of the critical issues in game based learning which are particularly resonant with our project, an introduction to the game and its development, and assignments which have proven effective in engaging students at creative and critical levels.

Games, Learning, and Caveats

There are many reasons to be optimistic regarding the outlook for digital pedagogies and the ancient Mediterranean. Many commercial video games offer adventures set in worlds informed by ancient history and myth, from the spectacularly visual *Assassin's Creed* to more rudimentary but enduringly popular games for Roman warfare, epic quests, and the construction of civilizations (McCall 2016; Christesen and Machado 2010; Bembeneck 2013; Graham 2014, 2017; Sabin 2012; Ghita and Andrikopoulos 2009). Archaeogames let players assume an archaeological avatar, experience challenge and rewards in ancient sites, and engage in survey and excavation (Mol et al. 2017; Reinhard 2018). These games intersect with virtual heritage projects in which participants experience 'materiality' in context and can interact with built environments, including Virtual Calakmul

shinydolphin.com/; Digital Projects Specialist, Joanna Mundy, Emory University; Leigh Cole Furrh and Alex Jester, Research Assistants; Craig Brasco, Visual Interface Collaborator, Kennesaw State University; Philip Kiernan, Outreach and Consultant, Kennesaw State University. Emory's Center for Digital Scholarship has generously hosted the game and the network and GIS analysis of Samothrace's cult http://digitalscholarship.emory.edu/; special thanks are due as well to Sara Palmer, Digital Text Specialist at Emory ECDS, and Michael Page, Geographer in Emory's Department of Environmental Sciences.

(Champion 2011: 4), Çatalhöyök, (Morgan 2009), and Rome at the pinnacle of its urban development (Guidi et al 2007). Such games and projects offer a meaningful pathway to engaging students with digitally enabled research in archaeology, epigraphy, and Geographic Information Systems, with issues of site preservation and conservation and the empowerment of descent communities as well as the world of classical reception (Champion 2011: 1-8; Cook Inlet Tribal Council 2017; Gordon 2017; Lowe 2009; Marshall 2019; McAuley 2019). Games devoted to the ancient Mediterranean are a subset of the larger category of historical games, whose potential for instruction has been a focus of conversation since the 1980s (Chapman 2016; Chapman, Foka and Westin 2017; Champion 2011; Elliott and Kappel 2013; Kee and Graham 2014; McCall 2011; Uricchio 2005). Those historical games, in turn, are a subsection of the vast conversation about games as media for instruction in business, neuroscience, information science, mechanics, sports training, and warfare. (Prensky 2001; McCall 2016; Gee 2007; Perla and McGrady 2011). The argument for their efficacy arises from the concept of serious games, foundational to which is the transferability of skills from game to real world (de Freitas 2018). The ancient environment of *Sailing with the Gods* is comprised of the sealanes of the Mediterranean, from the Greek mainland into the Black Sea, the Asia Minor coast and Egypt, rather than a single built context; players operate in the spaces in between which constituted the 'corrupting sea,' the arena for the formation of identity, fortune, and reputation in both legendary and historical Greek contexts.

Despite the difficulty of measuring effectiveness across multiple disciplines, a consensus is clear: some games, for some topics, offer measurable pedagogical benefits (Boyle et al. 2016; Van Eck 2015; de Freitas 2018; Arnab et al. 2012; Vandercruysse et al. 2012). The effects are measured at perceptual, cognitive, behavioral, affective, and motivational levels. Patterns that recur across these studies suggest the motivating force of a 'fun' activity (Prensky 2001, Arnab et al. 2012), also characterized as 'increased engagement' (Arya et al. 2012, Vandercruysse et al. 2012) and a response to problems with concentration span (Bellotti et al. 2011), all of which capitalize on the student experience of game culture (Burn 2016). Learning in game contexts is active rather than passive, exploratory and discovery based: the games promote situated learning, constructivist learning and problem solving (Burn 2016, Bellotti et al. 2011, Arnab et al. 2012, Van Eck 2006, 2015). These factors encourage the emergence of a community of players whose experiences reinforce each other. Of particular value for historical pedagogy is the extent to which gameplay encourages players to think critically and experience the outcome of their choices; their decision making mirrors the role of human agency in the working out of history, and so deepens the students' empathy toward their historical subjects (Takeuchi and Vaala 2014; Arnab et al. 2012). The games also foreground the confluence of multiple factors, including environment, economy and communication systems appropriate for the time, as well as the role of collaboration in the construction of archaeological knowledge (Arya et a.l 2012; McCall 2016). This capacity to foster an epistemological critique is a significant counter to the instrumental model which approaches gaming as a pedagogical trick, 'fairy dust,' to make dull material interesting (Burn 2016; Van Eck 2006, 2015; Champion 2011). This model overlooks the cultural power of games as an expression of what makes us human, fundamental from Huizinga onward. That cultural forcefulness is precisely what makes games effective, and shapes the cognitive scaffolding of players (Van Eck 2006, 2015). That scaffolding extends beyond the alluring aesthetics of past worlds as recreated on the console: games rely on risk and reward, and create meaning by drawing players into the concerns of the ancient people they study (Burn 2016; Bellotti et al. 2011; Gillings 2002). Active learning is stimulated in *Sailing with the Gods* through the absence of maps: players learn their route through exploration of the simulated topography, by hiring navigators, and reminders to follow the rising of the sun if they wish to head east. The openness of the quest line, which lets players choose whether to chase after fame or try to replicate the search for the golden fleece, encourages continual weighing of options in light of emergent opportunities. And the goal of the game, as per Van Eck's proposal, is to engage players not simply with a mythic quest, but the social structures of collaboration and cooperation that aligned the goals of ancient city states with those of their ambitious and highly mobile citizens.

There are, however, multiple critiques and caveats about gaming for history—indeed, the pedagogical revolution Pensky anticipated in 2001 has largely failed to materialize (McCall 2016: Van Eck 2006, 2015; Boyle et al. 2016; Takeuchi and Vaala 2014; Arnab et al. 2012). Notable among the impediments are the difficulty and time required to integrate games into a curriculum, particularly commercial, off the shelf games. Egenfeldt-Nielsen (2007) is an illustrative case study. Commercially produced games neither fit into his one-hour class period, nor suited a single-teacher classroom; they included excessive amounts of material irrelevant to his pedagogical aims, and ultimately proved disruptive. His response was to build his own game, *Global Conflict: Palestine*, in which the player becomes a journalist covering the Arab-Israeli conflict. Games to be used in education should optimally be designed with pedagogy in mind, but such attempts are still in their infancy (Bellotti et al. 2011). Van Eck's alternative is to have students themselves build games from the ground up, a perspective that reflects the 'maker movement' which emphasizes creation rather than consumption (2006, 2015). This approach has proven effective in our pedagogical strategies in *Sailing with the Gods*, engaging students at the level of data research on the one hand, illustrations on the other, depending upon the courses in which they are enrolled and their level of skill. Commercially based archaeogames, such as the Lara Croft franchise, compound these challenges with the tendency to model looting, destructive and illegal archaeological practices, the usurpation of indigenous pasts, trivialization of culturally sensitive material, and a masking of the unevenness and partiality of complex archaeological materials (Mol et al 2017, and Champion 2016.). The burden remains on instructors to construct clear goals for game play, establish patterns of feedback and debriefing, and construct meaningful bridges between course content and gameplay. Key among the latter are critical assignments that encourage students to compare good and bad archaeological methodology, and to contrast traditional historiographic narratives with those that emerge in the course of gameplay.

The issue of inaccuracy underlies a larger theoretical matter for games set in the past. Two different models of history clash in the critique of video games: history as the past, and history as a text. The first views prioritizes curated facts and historical accuracy, the Rankean 'wie es eigentlich gewesen ist'; the second foregrounds the activity of historiography as a post-modern representation, a narrative evolving continually in the work of historiographers (Uricchio 2005; Chapman, Foka and Westin 2017). A game to fit the first model of history would consist of a something like a quiz show, which would reinforce but not challenge the status quo (Elliott and Kappel 2013). Interactive video games, on the other hand, open the door to contrafactual scenarios and the ongoing process of historical inscription that is the heart of post structural historiography (Champion 2016). Player choices made in the game may mean that Napoleon wins at Waterloo; in god-games, such as *Civilization*, the mismanagement of water and irrigation could alter the course of Roman history. This open ended, agency driven quality distinguishes the ludic experience from a computer modeled experience of the past, which may lead a visitor through a fixed series of events (Champion 2014, 2016). The agency of the player writes the narrative of history and becomes a direct analogue to the choices of the historian who selects facts, pursues a hypothesis, and presents an argument. History is conceptualized as a shared process stretching over multiple forms, rather than a collection of details and established grand narratives. Uricchio notes that the poststructural historical turn was approximately contemporary with the emergence of hypertext and games: the hypertextual foundation of games corresponds to the demands to investigate historical possibility, empower the agency of the user, and offer access to a multi-valent, bottom-up writing of history. Accuracy of detail is less a generator of meaning than the intricate intersection of circumstance, conditions and contingency (Uricchio 2005; McCall 2016). Game designers may even introduce inaccuracies in order to reinforce play, as over-emphasis on factuality may impede playability (Elliott and

Kapell 2013). Sid Meier, who developed *Civilizations*, notes that one needs just the right amount of reality: success is shaped as much by what is not included as by what is, and too much information makes a game too arcane or controversial for its own good (Uricchio 2005). *Sailing with the Gods* takes up the open-ended model of history: players engage in mythopoiesis in a historical sea, reshaping the story of the voyage through their own responses to multiple, complex inputs from local histories, social structures, and Mediterranean ecosystems. This engages them with the non-canonical nature of myth, as a cultural product simultaneously ad hoc and informed by the narratological webwork that connected Greeks to each other and to a shared cultural identity.

Sailing with the Gods: animating a research question

Sailing with the Gods was not conceptualized from the beginning as a pedagogical tool, but emerged from a scholarly analysis of the mystery cult of the Great Gods of Samothrace, which uniquely promised its initiates safety from disasters at sea. The hypothesis is that this worked: the human social networks generated by grants of theoria and proxenia provided a civic, pragmatic, experiential counterpart to the mystical promises. Those promises are expressed in a range of forms, from the Nike on her prow, to the pinakes that crowded the sanctuary, to the rumors of the ithyphallic and apotropaic form of the gods of the rites. Ritual forms invoked in the rites include curses, apotropaia, votives, prayer, initiation, epiphany, prophecy, and apotheosis. These translate in myth into the interventions of a range of gods, including Dioskouroi, Glaukos, and Leukothea. These symbolic articulations of hopes for success had civic counterparts in the island's festivals, where proxenia was added to theoria for individuals who came especially from Asia Minor, the islands, and the Black Sea. These grants are preserved on a series of block grants, dated largely to the 2nd c BC. Inscriptions from within the city walls point to an association with the tradition maritime benefits of *proxenia*, including *eisploun, ekploun, asylia, ateleia*. These positioned the promises relevant to the rites in the civic guarantees of the island, and the festivals at which those benefits were granted. Those festivals facilitated the personal ties, recognition, and information flow that was the heart of anthropogenic safety around the Mediterranean (Blakely 2018).

The inscriptions also provide material for both GIS and network analysis, two approaches for which Django, an open-source, Python-based web framework, provided a common platform. We have nearly 1000 individuals and some 109 cities. The network analyses have yielded intriguing outcomes. They highlight the extent to which some cities seem to have used their connection to Samothrace to build local identities but not network connections; they also identify sites such as Thasos and Aigai which play roles as connectors between neighborhood clusters. The analysis overall yields an image of a highly clustered, scale free network, suggesting non-random patterns of formation that may open the door to modeling backwards how the system emerged. The latter is a point of particular interest, as the promises of safe sailing are well in place in the sixth and fifth centuries BC, but the abundant epigraphic record is dated largely to the 2nd and 1st centuries BC (Blakely 2016a).

Alluring as these outcomes were, they are encumbered by all the limitations that attend network analyses in ancient contexts. We are attempting quantitative and statistical analyses of data which is partially preserved: the site's lime kilns may have taken materials that would yield entirely different outcomes. The Cartesian coordinates of some of our cities are matters of debate, with some having three to five different possibilities. And most critically, the exact working out of proxenia on the ground as a strategy for maritime safety relies on factors beyond the privileges specified on the stones. They rely on the capacity for recognition, and the multiplier effect of numerous social networks coming together in the context of the festivals. There is, perhaps most pointedly, no capacity to read individual human agency, the motivations and outcomes for those who invest in the Samothracian trek. A gaming application offered an attractive route forward, because of the principle of serious games to model behavior that transfers from the abstract and imaginary to the real, and the potential for platforms like Unity, a cross-platform game engine for 2D and 3D video games, to integrate the physical world of land and seascapes, the social history of civic institutions, and the cultural traditions symbol and narrative into a single interactive experience. The bottom-up approach to history that underlies this network analysis, moreover, aligns with the postmodern framework of history that emerges from gaming. While recipients of proxenia and theoria would be notable citizens, all of them fly far below the radar of traditional historiography; virtually none of the names from the epigraphic record of Samothrace appear elsewhere in the historical record. The ultimate goal of the game is the generation of data regarding player choices made in a networked sea, to complement our analysis based on the fragmentary ancient record.

In order to move from network analysis to effective game play, we began by adding a quest line, a narrative that couples characters with objectives that engage the player's imagination and excitement: we found this in the *Argonautica* of Apollonius of Rhodes. We chose the Argonauts for several reasons. Some versions of the epic include the detail that Jason and his crew stopped in at the island for initiation, and the pathway the Argo pursues through the northern Aegean sea and up the Bosporos coincides significantly with the Northern Aegean maritime route, with roots back into the Bronze age, on which Samothrace was a significant port of call (Papageorghiou 2009). Individual Argonauts figured in the political interactions among the Hellenic cities of the Black Sea: Herodotus claimed that the Lemnians who claimed land on Taygetos persuaded the Spartans to support them based on claims of descent from the Argonauts (Herodotus 4.145.2; Braund 1996). The Hellenistic period was the great floruit for the Samothracian rites, and the time when Apollonius of Rhodes wrote his Argonautica for the Ptolemaic kings who were its great patrons. The epic and the rites are both cultural responses to maritime risk, the *Argonautica* as a paradigmatic first voyage, the rites as ritual insurance (Jackson 1997). The ritual vocabulary of maritime safety figures large in the epic as well, as the Argonauts establish landmark altars, calm the winds, and leave a trail of sacred anchors and buried seamen behind them (Blakely 2016b).

Figure 1: Fog fades into the screen at ~15 km

Figure 2: Night sky: note lines indicating constellations. White streak is a visual indication of moving winds.

Classroom approaches to the *Argonautica* often focus on the intense emotional and ethical crises of Jason and Medea: our interest is the journey, and accordingly significant attention has been dedicated to the accuracy and visual engagement of the process of sailing. Our platform is Unity, which supports the integration of landscape, social structures and cultural narratives. Our topography is scaled and georeferenced using satellite imaging data, accurate to a resolution of roughly 1km.[2] The simulation, in order to come closest to the visual experience of an ancient sailor, replicates the curvature of the earth as viewed from a height of 10 meters; the height provides a bird's eye view of the ship in context. A fog fades in at ~15 km, to augment the sense of fading visual access for objects at a distance (Figure 1). The stars were programmed using a predictive model of axial precession determined by NASA, so that they are in the appropriate positions for the year 200 BC (Figure 2).[3] The stars are presented in a series of nested celestial spheres that rotate around the ship, enabling a 'mariner-centric' simulation. Similarly focused on the sailor's perspective are our concerns for the embodied reality of sailing: incorporated into the game's algorithms are requirements for food and water based on data from NASA and the National Academies Institute of Medicine, slightly increased to account for the demands of rowing and the draining effects of direct sunlight.[4] Our earliest play-throughs were punctuated by the frequent deaths of sailors from dehydration, high-

---

[2] The simulation uses GTOPO30 elevation data; the dataset has a grid space of approximately 1 km. One unit of Unity world space is the equivalent to 1193.920898m in the processed, projected dem raster. A subset of the GTOPO30 dataset is used to reference only the Mediterranean area and project into the EPSG:32634 standard. The projection used is for visual reference only. For accurate placement of cities, the origin of the unity world space (0,0,0) is set to match the south-western corner of the GTOPO30 subset; latitude and longitude were taken from Pleiades https://pleiades.stoa.org/

[3] The constellation maps were taken from NASA's Tycho Catalog Starmap and Deep Star Maps The formula for axial precession used is referenced from: N. Capitaine et al 2003. Expressions for IAU 2000 precession quantities. P581. DOI: 10.1051/0004-6361:20031539.

[4] The daily intake of food is roughly .71 kg per day once the astronauts' food packaging is removed. The IOM's recommended daily intake of water for the average male is 3.71l per day (also 3.71kg). Because the crewmembers are onboard a ship and performing strenuous physical activities such as rowing and being subject to direct sunlight, we have raised the average to an even 5l of water per day per crew member. NASA Food intake – http://www.nasa.gov/vision/earth/everydaylife/jamestown-needs-fs.html National Academies Institute of Medicine (IOM) water intake –http://www.nationalacademies.org/hmd/~/media/Files/Activity%20Files/Nutrition/DRIs/DRI_Electrolytes_Water.pdf

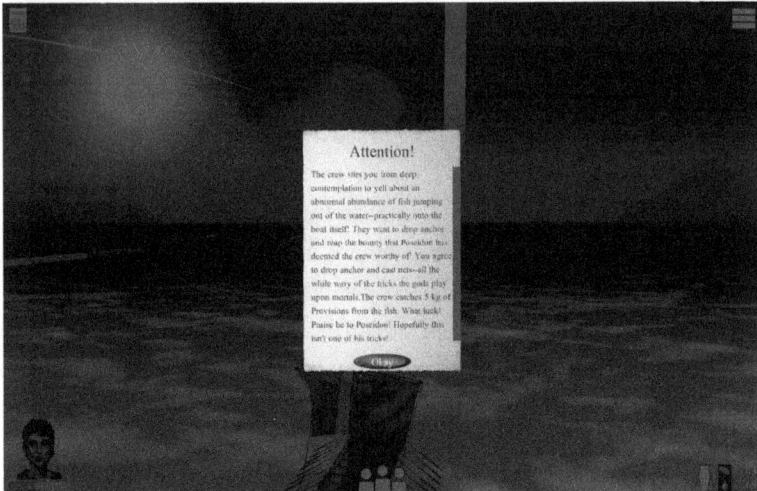

Figure 3: Random event: an unlooked for abundance of fish

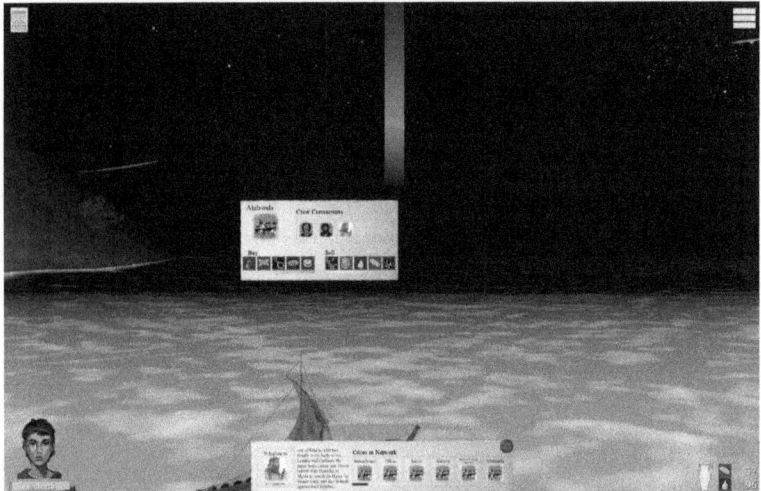

Figure 4: The beam of light marks the direction in which Alabanda lies; the popups show how players can get information about any city in the proxenic network of the current crew. The pop-up shows that three current crewmembers have proxenia grants from this town, as well as the objects most suitable for buying and selling. Details on one of these three, Polyphemus, appear at the bottom of the screen: we see his biography, and all the cities he has in his network.

lighting a critical need for water stops that lies far beyond the epic poet's interests. Random events reflect the intersection of human and divine agency (Figure 3).

Equally central to our research interests is the impact of the social networks that were formalized in proxeny grants. Proxenia was a complex institution, embracing a history of service to the city, an eagerness to perform acts of goodwill in the future, and a symbolic tie to the deep history of xenia as a Greek cultural strategy: it was part of the Hellenistic webwork that bound geospatially separated entities to each other. Its effectiveness derived from the extent to which it was both widespread and stereotypic. Because of this, the database for proxeny decrees needed to be expanded beyond Samothracian proxeny alone. The sheer scope of the data, and the principles of effective game design, stipulated the selection of a meaningful representative sample. We have in this game a total of 249 cities and 130 individuals, informed by the proxeny decrees as preserved on commemorative lists (Mack 2015). A narrative clip has been written for each city, which offers historical and mythic details of the place. Each individual has a brief biography based on ancient sources, and brings with him the proxenic network of his home city (Figure 4). A game, moreover, in order to model choices based on proxenia, must find a way to express these complex notions numerically. Rob Bryant, the lead designer for the game, designed an algorithm that quantifies the value of proxenia—that held by Jason, and that held by all members of the crew—as the ship moves from one port to another. This is quantified in the form of freedom from port taxes, more generous flows of information in port, and better market prices for buying and selling. These, in turn, translate directly into the player's ability to 'level up' from goatherd to god. Wealth on board enables the replacement of men who die or abandon ship; it allows the creation of monuments that elevate the fame of the dedicator; and it lets players purchase food, water and wages for sailors, staving off the starvation, dehydration and revolt that would imperil the voyage.

To play the game, players assume Jason as their avatar, and select 20 crew members from the epic, from one Samothracian block grant, or from other cities with active proxenic networks. These

Figure 5: the pop-ups inform players of the dangers of piracy, with details drawn from ancient historical sources.

crewmembers will need to be replaced during the course of play, due to casualties arising from outbreaks of disease on board, pirate fights, starvation, or dehydration (Figure 5) All individuals, whether drawn from the epic or from the epigraphic record, come from historical Greek cities and bring with them their city's proxenic networks. Thus a Euboian named Canthus came from the *Argonautica,* and brings to the ship all the network connections of the Aitolian league, including the cities Histiaia, Karthaia, and Chalkis. The object of the game is to get famous, get rich, and not die—it is, most notably, not to replicate the mythical route of the Argo, but a free experimentation with historical strategies for success in an ancient sea. What the players share with the ancient heroes is an ambition for *kleos*, which in this game is presented as 'clout.' Clout may be accumulated through factors that resonate with epic heroes, including the military excellence shown by overcoming pirates, the creation of memorials, and the combined clout of one's colleagues; it may also come incrementally, by supplying the biological imperatives of food, water and rest for the crew, and overcoming the risks of storms and disease. The focus on making profit in port is appropriate for the well-recognized economic functions of proxenia (Engen 2013: 140-182).

Pedagogical goals and classroom strategies

Three pedagogical goals have emerged as natural foci for *Sailing with the Gods*. The first is that myth is not canon: the experience of game play foregrounds the distinction between the literary texts we tend to favor in mythology courses and myth as a phenomenon that does cultural work in Cartesian spaces. The second is the fresh light cast on Apollonius' text itself, specifically his fascination with geography, harbors, and the periplus tradition, and the existence of those resources on the shelves of Alexandria's library. The third is the concept of human geographies: students come out of the game with a view of the ancient Mediterranean structured as much by stories told and contracts agreed to as a purely geospatial configuration. The achievement of those goals relies on pre-game presentations as well as the assignments themselves. These begin with the story of the game's invention. For my students, who are only rarely majors in Classics, Art History or History, this bridge between their coursework and faculty research deepens their sense of the value of the ancient materials. The focus on social networks highlights a human impulse for connectivity that connect their world with antiquity. A second component is an introduction to the multiplicity of known ancient Argonauticas, from Skytobrachion to Valerius Flaccus. This helps minimize the desire to simply repeat Jason's route in the process of game play; it also makes that game play resonate with the historical patterns of myth's responsiveness to authorial, historical, and local political needs. A third element is Apollonius as an author, fascinated with geography, ktiseis, and harbors, and the presence of these resources on the shelves at Alexandria. This helps the students focus on the voyage as much as the drama of the text, and clarifies the cultural discussion about maritime strategies as the canvas on which the epic is painted. The focus on clout, rather than Medea or the fleece, concludes the presentation. This encourages comparison, on the one hand, between Jason and heroes such as Achilles and Odysseus; it also helps the students read the epic heroes as idealizations of ordinary men who thrived in an ancient networked sea—and whose daily experiences the students mirror as they barter in port, seek cooperative shores, and strive to level up in a sea filled with risk.

BOX 1

Description: Apollonius of Rhodes' Argonauts build on a long tradition of local myths scattered around the Mediterranean. As the heroes landed in strange places, they brought their own accounts of who they were and where they came from, and if they were good travelers they would leave every new city with a deeper understanding of the local narratives that defined the place.

Goals:
- Build on your understanding of myth in the landscape.
- Introduction to online scholarly sources.

Select three different locations from Apollonius' Argonautica – cities, cult sites, rivers.

Find one mythic tale other than the Argonautica, using BNJ, BNP, Perseus, PECS, LIMC, Pleiades.

For each myth, **provide**:
- reference in Apollonius of Rhodes' text.
- Brief description of the story.
- Author, work, and specific reference.
- URL.

Assignments: data gathering, maker culture, and historical comparison

Three different types of assignments offer pathways to these learning outcomes: data gathering, creative artwork and data analysis, and a debriefing and feedback that engages historical materials. The first responds to a significant challenge in our game: the sheer scale of our geospatial reach. Unlike archaeogames set in a single archaeological site, the seas on which Samothracian initiates sail stretched from Alexandria to Kyzikos to Rome. This means there is an endless amount of potentially relevant material for cities, landscapes, and personalities. This problem has presented a productive pedagogical route, engaging students with data gathering and with the research engines – including *Brill's New Jacoby*, *Brill's New Pauly*, Perseus, the *Princeton Encyclopedia of Classical Sites*, LIMC and Pleiades – which enable them to find materials beyond those assigned in class (see Box 1). The assignment helps students experience the link between myths and geospaces, with all of their political and economic concerns; they also often encounter local elaborations of characters whom they encountered in other readings, further pressing the edges of a purely literary understanding of myth.

Craig Brasco at Kennesaw State University has generously collaborated with us to enable the creation of a second assignment, one oriented toward creative engagement in the spirit of the maker movement. Students enrolled in Kennesaw's new BFA program in digital animation were introduced to the game, offered a brief verbal snapshot of each crewmember, and a limited number of images as inspiration. These images, frequently coins or ceramics, materialize for the students the use of elements from the natural and mythological worlds as second level signifiers in an ancient context. These artifacts rendered the purchase of goods and the drinking of wine into expressions of identity and cultural memory, and so offer fresh light on the work the students do in crafting the cultural imaginarium of the digital age. Their results contribute both to the game and to the students' emerging portfolios; students retain the rights for their images, and are credited on the game's website. Among the results are (Figure 6)

Figure 6: Rhodian sailor, by Ren Nesper, Kennesaw State University

Figure 7: Erginus, by Paola Paniagua, Kennesaw State University

Figure 8: Tiphys, by Ren Nesper, Kennesaw State University

Figure 9: Telamon of Aegina, by Julee Davis, Kennesaw State University

the generic "Rhodian sailor", on whose neck is a tattoo of the canting rose of Rhodian coins. Erginus, identified as the semi-divine, Milesian son of Poseidon, bears a face tattoo of his father's iconic trident and a shield invoking the Mycenaean stratum of his hometown as well as the creatures of his father's realm (Figure 7). Tiphys is identified in our prompt as the pilot of the Argo, whose participation was assured by Athena herself, and who always observed the wind and stars. The result is a red-eyed young man hallucinating constellations (Figure 8). Telamon of Aegina, who berated Jason as too intimidated to compete with the greatest of heroes, stretches his muscle-bound arm behind his head to show his own tattoo, drawn from his island's coins (Figure 9).

This maker movement sensibility extends as well to students and colleagues in programming and network analysis courses. The entire set of data for network analysis, as well as the most recent version of the game, are available for download on GITHUB. One of my Emory colleagues, Jeremy Jacobsen of Emory's Institute for Quantitative Theory and Methods, has made a python-based network analysis of the Samothracian geospatial and epigraphic record one option for his students' final projects in Data Science Computing 250, a course focused on inferential techniques not found in introductory statistics courses, including clustering and neural networks. The data set is ideal for the course's orientation to practice with real-world data sets and creative projects. The outcomes have been an intriguing complement to our Gephi-based analyses, and we welcome analogous collaborations in the future, particularly from courses focused on programming for game design.

A third type of assignment is based on critical comparison with historical materials covered in class—a feedback loop that challenges the students who are neither in creative nor computer courses to make their own synthesis of the gaming and historical worlds. This kind of post-play debriefing supports the transfer of game experience to real world experience, on analogy with the principles which underlie the use of serious games in military, business and medical training (Perla and McGrady 2011; Loh et al 2015: Crookall 2010). In *Shipwrecks, Pirates and Palaces,* an undergraduate course onthe

> BOX 2
>
> **Read** the introduction to the game on the website.
>
> As you play, **note**:
> - the crew you select (each has his own skills, place of origin, and story).
> - the impact of various costs and risks: financial (the cost of doing business), physical (rowers need food and water), sociogenic (pirates can show up).
> - the ways you can increase your clout, the game's analogy for ancient kleos.
> - the algorithm for calculating different costs in port based on the proxenia of your crew.
>
> **Offer** four observations that connect you game experience to (1) a text (2) a concept (3) a case study we have covered in class. Be as specific as possible. If you encounter pirates, you may refer to the damage attributed to the sea peoples in the inscription of Ramses III at Medinet Habu; if you make a significant profit selling grain, you could cite the profit made by the koina of farmers and shippers at Lindos as a comparison.

archaeology of the ancient economy, the game has become the focus of one week's writing assignment. By the last four weeks of the term, students in this course have been introduced to civic institutions, port taxes, piracy, trade routes, and have drawn on mythic as well as historical texts that address landscape and resources, types of exchange and the potential for profit. The students are instructed to select their crew wisely, to play for at least 20 minutes, and to identify four analogies between experiences in the game and materials encountered in the course. (See Box 2) What is significant in the assignment is that they are not simply hunting for specific economic factors embedded in the game: they are forming analogies with course materials, using the categories of evidence with which they have become familiar over the course of the term. One student noted that when her ship was attacked by pirates, they stole gold and provisions and kidnapped two men, comparable to the pirates in Andokides' *Mysteries* 138: "The sea was infested with triremes and pirates, who took many a traveler prisoner." A second student found that pirates took half his commodities, killed two crewmembers, and took three additional men to sell into slavery—the latter consistent with Lysias 1.3 *Against Agoratos*. A third described her ship coming upon a sinking vessel and taking whatever they could find, and noted that this was consistent with the law of *res communis*, which made materials lost at sea into common property.

The same principle informs assignments for mythology and literature courses. After students read the *Argonautica,* they receive an assignment to play the game for at least 20 minutes, and then return to their texts to identify and discuss a 5-10 line passage they read differently after playing. Their responses foreground two key ideas. The first reflects the deepened empathy with historical characters described by Takeuchi and Vaala 2014) and Arnab et al (2012). Students discuss the difficulty of sailing in an ancient sea, a realization that reshapes their analysis of Jason's reluctance and frequent faintness of heart regarding the toils of the voyage. The second demonstrates a new approach to the text as a response to its environment. These students report that after bobbing about on a ship at night, with nothing but the light of the stars above and the dark water beneath them, the song of Orpheus about the formation of the universe—focusing on the earth, heaven and sea, the celestial paths of the stars and moon, and the role of the daughter of Ocean—made sense to them, as a response to the lived environment (*Argonautica* 1: 495-512; cf. Arya et al 2012; McCall 2016). The use of the game alongside a canonical text also enables an experiential understanding of the postmodern approach to literature that is particularly appropriate for a course foregrounding the relationship between mythic traditions—evolving and adaptive—and canonical literary texts. Gee articulated the alignment between the open-endedness of gameplay and the instability of canonical texts (2007), analogous to the suitability of gameplay to post structural, polyvocal historiography. The multiple versions of the *Argonautica* offer a historical exemplum of such multiple subjectivities.

The gameplay extends that historical process of evolution and makes tangible the contingency of the narrative as it unfolds in different directions in response to player choice. The students' agency as they play maps onto the choices made by ancient authors and patrons who adapted Jason and his crew to their own needs. Those changes, moreover, are driven by the players' pursuit of their clout, the drive to level-up that engages them with both the kleos-hungry heroes of the Greek mythic tradition and the real potential for economic success in the Hellenistic world. The engagement with kleos and proxenia represents the 'aligning of subjectivities' through which Uricchio proposes that historical games connect the player with past worlds at the levels of knowledge, motives, and perceptual horizons. The students transfer specific concepts of maritime risk and an articulate landscape back to their reading; more profoundly, they may see a deeper level of themselves in the text. This is a far deeper engagement than that afforded by the instrumental model that would make game simply the delivery system for historical details.

Conclusions and caveats

Our goal in creating this game was the generation of sufficient amounts of data to yield meaningful materials for more network and geospatial analysis. That goal lies ahead of us, as the amount of data we will need relies on negotiating the divide between games assigned in class and games played by choice—a longstanding divide familiar in the scholarship on game-based learning. Our pedagogical applications to date, however, have yielded unanticipated insights. The first is the patience and even eagerness on the part of the students to work with a buggy, imperfect game in process. This opens the possibility of engaging them with the process itself at the level of research and creativity as well as feedback: they contribute in meaningful ways to the whole, and the process intensifies their critical and synthetic lenses on ancient material. The second is a confirmation of the gestalt approach to gaming, one that affirms the analytical and cognitive value of games even when they are not visually sophisticated. The earliest version of our game were cartoonish and low on illusionistic accuracy, and we still lack the animation of figures that could help close the gap in sophistication between our algorithms and the user interface. Even in this cartoonish form, however, the outcomes suggest the potential to give the students an experiential perspective on the environment in which the text functioned. The third is the great value of assignments that send the students back from the game for a second look at the primary materials on which our discipline is founded. Concerns that the games are more of a distraction than a tool, or that their inaccuracy is counterproductive, find an answer in engaging the students in the analysis of games vis-à-vis the ancient materials, their interest in which, after all, is what brings them to our courses. Most clear is that we have only begun to scratch the surface, and that we will learn as much as our students in the years ahead.

Works Cited

Arnab, S, R. Berta, J. Earp, M. Usart.
2012 Framing the Adoption of Serious Games in Formal Education *Electronic Journal of e-Learning* 10(2): 159-171.

Arya, A., P. Hartwick, S. Graham, N. Nowlan.
2012 Collaborating through Space and Time in Educational Virtual Environments: 3 Case Studies. *Journal of Interactive Technology and Pedagogy*, 2: 1-28. http://jipt.commons.gc.cuny.edu/collaborating-through-space-and-time-in-educational-virtual-environments-2-case-studies/.

Bellotti, F., M. Ott, S. Arnab, R. Berta, S. de Freitas, K. Kiili, A. De Gloria.
2011 Designing Serious Games for education: from Pedagogical principles to Game Mechanisms. In *Proceedings of the 5th European Conference on Games Based Learning, The National and Kapodistrian University of Athens, Greece, 20-21 October 2011*, edited by D. Gouscos and M. Meimaris, pp. 26-34. Academic Publishing, Reading, .

Bembeneck, E.J.
2013 Phantasms of Rome: Video Games and Cultural Identity. In *Playing with the Past*, edited by M.W. Kapell and A.B.R. Elliott, pp. 77-90. London, Bloomsbury Academic.

Blakely, S.
2016a Beyond Braudel: Network models and a Samothracian Seascape. In *Across the Corrupting Sea*, edited by L. Mazurek and C. Concannon, pp. 17-38. London, Ashate.

Blakely, S.
2016b: 'Maritime Risk and Ritual Responses: Sailing with the Gods in the Ancient Mediterranean'. In *The Sea in History – The Ancient World*, edited by C. Buchet and P. de Souza, pp. 362-379. Woodbridge, The Boydell Press.

Blakely, S.
2018: 'Sailing with the Gods: Serious Games in an Ancient Sea', *Thersites* 7: 108-146.

Boyle, E. A., T. Hainey, T.M. Connolly, G. Gray, J. Earp, M. Ott, T. Lim, M. Ninaus, C. Ribeiro, J. Pereira.
  2016 An update to the systematic literature review of empirical evidence of the impacts and outcomes of computer games and serious games. *Computers & Education* 94: 178-192.

Braund, D.
  1996 The Historical function of Myths in the Cities of the Eastern Black Sea Coast. In *Sur les traces des Argonautes* edited by O. Lordkipanide and P. Lévêque, pp. 11-19. Paris, Les Belles lettres.

Bresson, A.
  2015 *The Making of the Ancient Greek Economy: Institutions, Markets, and Growth in the City-States*, tr. S. Rendall. Princeton.

Burn, A.
  2016 Liber Ludens: Games, Play and Learning. In *The SAGE Handbook of E-learning Research* edited by C. Haythornthwaite, R. Andrews, J. Fransman, & E. M. Meyers. Second Edition, pp. 127-151, Los Angeles, Sage.

Champion, E.
  2011 *Playing with the Past*. London, Springer.

Champion, E.
  2016 Entertaining the similarities and distinctions between serious games and virtual heritage projects. *Entertainment Computing* 14: 62-74.

Champion, E.
  2014 History and Cultural Heritage in Virtual Environments. In *The Oxford Handbook of Virtuality*, edited by M.. Grimshaw, pp. 269-283. Oxford, Oxford University Press.

Chapman, A.
  2016 *Digital Games as History: How Videogames Represent the Past and Offer Access to Historical Practice*. London, Routledge.

Chapman, A., A. Foka and J. Westin.
  2017 Introduction: What is historical game studies? *Rethinking History: The Journal of Theory and* Practice 21(3): 358-371.

Christesen, P. and D. Machado.
  2010 Video Games and Classical Antiquity. *Classical World* 104(1): 107-110.

Cook Inlet Tribal Council.
2017 Storytelling for the Next Generation: How a non-profit in Alaska harnessed the power of video games to share and celebrate cultures. In *The Interactive Past: Archaeology, Heritage & Video Games*, edited by A. Mol, C.E. Ariese-Vandemeulebroucke, K.H.J. Boom, and A. Politopoulos, pp. 21-32. Leiden, Brill.

Crookall, D.
2010 Serious Games, Debriefing, and Simulation/Gaming as a Discipline. *Simulation & Gaming* 41(6): 898-920.

De Freitas, S.
2018 Are Games Effective Learning Tools? A Review of Educational Games. *Educational Technology & Society* 21(2): 74-84.

Egenfeldt-Nielsen, S.
2007 *Educational Potential of Computer Games*. London, Continuum.

Elliott, A.B.R. and M.W. Kapell.
2013 To Build a Past That Will "Stand the Test of Time" – Discovering Historical Facts, Assembling Historical Narratives. In *Playing with the Past: Digital games and the simulation of history*, edited by M.W. Kapell and A.B.R. Elliott, pp. 1-30. London, Bloomsbury Academic.

Engen, D.T.
2013 *Honor and Profit: Athenian Trade Policy and the Economy and Society of Greece, 415-307 B.C.E.* Ann Arbor, University of Michigan Press.

Gee, J.P.
2007 *What Video Games Have to Teach Us about Learning and Literacy*. Second edition. New York, Palgrave Macmillan

Ghita, C., and G. Andrikopoulos.
2009 Total War and Total Realism: A Battle for Antiquity in Computer Game History. In *Classics For All: Reworking Antiquity in Mass Culture* edited by D. Lowe and K. Shahabudin, pp. 109-126. Newcastle upon Tyne, Cambrdige Scholars.

Gillings, M.
2002: 'Virtual archaeologies and the hyper-real', in P. Fisher and D. Unwin (eds), *Virtual Reality In Geography*, London, 17-32.

Gordon, J.
2017 When Superman Smote Zeus: analysing violent deicide in popular culture. *Classical Receptions Journal* 9(2): 1-26.

Graham, S.
2014 Rolling Your Own: On Modding Commercial Games for Educational Goals. In *Pastplay: Teaching and Learning History with Technology* edited by K. Kee, pp. 214-254. Ann Arbor, University of Michigan Press.

Graham, S.
2017: 'On Games that Play Themselves: Agent based models, archaeogaming, and the useful deaths of digital Romans. In *The Interactive Past: Archaeology, Heritage & Video Games*, edited by A. Mol, C.E. Ariese-Vandemeulebroucke, K.H.J. Boom, and A. Politopoulos, pp. 122-132. Leiden, Brill.

Guidi, G. B. Frischer, J. Donno, M. Russo.
2007 Virtualizing ancient Imperial Rome: From Gismondi's physical model to a new virtual reality application. *International Journal of Digital Culture and Electronic Tourism* 1(2): 240-252.

Jackson, S.
1997 Argo: the first ship? *Rheinisches Museum für Philologie* 140(3/4): 249-257.

Kapell, M. W. and A.B.R. Elliott (eds).
2013 *Playing with the Past: Digital games and the simulation of history.* London.

Kee, K. and S. Graham.
2014 Teaching history in an age of pervasive computing: the case for games in the high school and undergraduate classroom. In *Pastplay: Teaching and Learning History with Technology* edited by K. Kee, pp. 337-366. Ann Arbor, University of Michigan Press.

Loh, C.S., Y. Sheng and D. Ifenthaler.
2015 Serious Games Analytics: Theoretical framework. In *Serious Games Analytics: Methodologies for Performance Measurement, Assessment, and Improvement* edited by C.S. Loh, Y. Sheng and D. Ifenthaler, pp. 3-30. New York, Springer.

Lowe, D.
2009 Playing with antiquity: Videogame receptions of the classical world. In *Classics For All: Reworking Antiquity in Mass Culture* edited by D. Lowe and K. Shahabudin, pp. 62-88. Newcastle upon Tyne, Cambrdige Scholars.

Mack, W.
2015 *Proxeny and Polis: Institutional Networks in the Ancient Greek World.* Oxford, Oxford University Press.

Marshall, C.W.
2019 Classical Reception and the Half-Elf ClericIn *Once and Future Antiquities in Science Fiction and Fantasy* edited by B.M. Rogers and B.E. Stevens, pp. 149-171. London, Bloomsbury Academic.

McAuley, A.
2019 The Divine Emperor in Virgil's *Aeneid* and the *Warhammer 40k* Universe. In *Once and Future Antiquities in Science Fiction and Fantasy* edited by B.M. Rogers and B.E. Stevens, pp. 183-195. London, Bloomsbury Academic.

McCall, J.
2011 *Gaming the Past: Using Video Games to Teach Secondary History.* New York, Routledge.

McCall, J.
2016 Teaching History with Digital Historical Games: An Introduction to the Field and Best Practices. *Simulation and Gaming* 47(4): 517-542.

McGonigal, J.
2011 *Reality is Broken: Why Games make Us Better and How They Can Change the World.* New York, Vintage.

Mol, A.A.A, E.C. Ariese-Vandemeuleroucke, K.H.J. Boom, A.Politopoulos.
2017 *The Interactive Past: Archaeology, Heritage & Video Games.* Leiden, Brill.

Morgan, C.L.
2009 (Re)Building Catalhöyük: Changing Virtual Reality in Archaeology. *Archaeologies: Journal of the World Archaeological Congress* 5: 468-487. https://doi.org/10.1007/s11759-009-9113-0

Papageorgiou, D.
2009 The Marine Environment and its Influence on Seafaring and Maritime Routes in the Prehistoric Aegean. *European Journal of Archaeology* 11(2-3): 199-222.

Perla, P.P. and E.D. McGrady.
2011 Why Wargaming Works. *Naval War College Review* 64(3): 111-130.

Prensky, M.
2001 *Digital Game Based Learning.* New York, McGraw Hill.

Reinhard, A.
2018 *Archaeogaming: An Introduction to Archaeology in and of Video Games.* New York, Berghahn Books.

Sabin, P.
2012 Ancient Warfare. In *Simulation War: Studying Conflict through Simulation Games* edited by P. Sabin, pp. 135-160. London, Bloomsbury Academic.

Takeuchi, L.M. and S. Vaala.
2014 *Level up Learning: A national survey on teaching with digital games.* New York, Games and Learning Publishing Council.

Uricchio, W.
2005 Simulation, History and Computer Games. In *Handbook of Computer Game Studies* edited by J. Raessens and J.H. Goldstein, pp. 327-338. Cambridge, MA, MIT Press..

Van Eck, R.N.
2006 Digital Game-Based Learning: It's Not Just the Digital Natives Who Are Restless. *EDUCAUSE Review* 41(2): 17-30.

Van Eck, R.
2015 Digital Game-Based Learning: Still Restless, After All These Years. *Educause Review* 50(6): 13-28.

Vandercruysse, S., M. Vandewaetere and G. Clarebout.
2012 Game based learning: A review on the effectiveness of educational games. In *Handbook of Research on Serious Games as Educational, Business, and Research Tools* edited by M.M. Cruz-Cunha, pp. 628-647. Hershey, PA, IGI Global.

Programming without Code: Teaching Classics and Computational Methods

Marie-Claire Beaulieu
Anthony Bucci

When considering what to cover in a joint class between Classics and Computer Science titled 'Computational Methods for the Humanities', the first and possibly most thorny question is to decide what technical skills need to be taught, and to what degree.[1] We started by asking ourselves what the job of a humanist entails. Our answer was that humanists interpret the world; they seek the significance behind cultural and societal facts. In their work, humanists have access to vast amounts of data in multiple formats such as literature, music, historical documents, material culture, spoken accounts, and so on. The question then became, in order to take advantage of technology to do this job, does a humanist need to learn coding?

Why would a humanist learn to code?

We decided that the class' main objective would be establishing fluency in working with data in digital form. Fluency includes knowing the steps involved in data analysis, typical formats and sizes of datasets, and techniques for data cleanup and manipulation, topics we chose to cover early in the course. Data fluency would in turn lead to modules geared toward the interpretation of a data-driven analysis, focusing on how to create and read plots, maps, and graphs. Throughout the term, the students would be exposed to current tools for working with data in the humanities, as well as important projects that

---

[1] Ramsay 2012 grapples with the same issue but chooses a different solution. See also Bonds 2014, Hall 2015, Kingsley 2016, Mahony and Pierazzo 2012.

release humanities data for research. Finally, and most importantly, the emphasis would be on the ability to formulate research questions and hypotheses, and to connect the results of the analysis back to the real world, as the humanities are grounded in real-world material and questions.

The fact that the class was offered as a joint enterprise between Classics and Computer Science allowed us to take advantage of the long-standing tradition of careful data collection in Classics, and to examine the transition between legacy data in analog form and its digital incarnation. We introduced our students to both the book and database formats of *L'Année Philologique*,[2] the Pleiades Gazetteer,[3] the *Lexicon of Greek Personal Names* (LGPN),[4] the *Corpus Vasorum Antiquorum* (CVA),[5] and the Mantis[6] database of coinage. Where applicable, searches of the same repositories in different forms revealed different, but not necessarily inferior, results. For instance, if performing a search for scholarship on the myth of Danae in the paper version of *L'Année Philologique*, one would look in the index for the names "Danae", "Acrisius", and "Perseus". It would also be wise to browse the "Greek literature > Greek Religion" sections so as to catch titles and topics such as "Women in Greek Mythology". While technically this search is possible in the online version of the database (fuzzy matches are enabled), the set of results found would no doubt be different, and complementary, to the results found in the analog version. This comparison, which we taught about during a class session held at our university library with the relevant books, brought home the point that understanding the significance of the data, its provenance, and its organization is the crucial start to any data analysis, in whatever form.

---

[2] http://www.brepols.net/Pages/BrowseBySeries.aspx?TreeSeries=APH-O.
[3] https://pleiades.stoa.org/.
[4] http://www.lgpn.ox.ac.uk/.
[5] https://www.cvaonline.org/cva/.
[6] http://numismatics.org/search/.

Learning to code presented as a problem of choosing among the many options.

Figure 1. Screen capture of a Google search for 'learn how to code'

In light of the quantity of material we could cover in preparing students to understand, work with, and interpret data, it was hard to see how there would be space to teach any programming concepts at all, let alone to do justice to the subject in a way benefiting students. While a search of the web for "learn how to code" (see Figure 1) suggests the process is relatively quick and painless, pedagogical reality suggests otherwise. Anecdotally, acquiring basic fluency in a programming language with limited prior exposure requires about the same effort as acquiring basic fluency in a natural language: 600-750 hours of dedicated study, according to the US Foreign Services Institute.[7] On top of this, a 2007 report by the Liberal Arts Computer Science Consortium suggests expanding introductory computer programming instruction to three introductory courses rather than two, reflecting the growth in material in the field over the last few decades.[8] What would we hope to convey to students in a single course

---

[7] See https://www.state.gov/m/fsi/sls/c78549.htm.
[8] Liberal Arts Computer Science Consortium 2007.

that was only partially dedicated to teaching computational methods, when the evidence suggests basic fluency requires significantly more effort?

Beyond the bounds of our own course, there are wider subject matter barriers to humanists learning to code. For instance, at Tufts University, the introductory Computer Science curriculum starts with two semesters of mandatory C++ classes. C++ is a programming language geared towards system programming and is unlikely to ever be of use to humanists. There are alternatives to undergraduate Computer Science curriculum coursework, such as a one-semester Python programming class or a six-week summer bootcamp in Python, Matlab, and R, but these can fill up quickly, especially during the regular semester, and the waiting list is typically very long. Additionally, programming instruction typically assumes a certain background knowledge in mathematics or STEM: problem sets will frequently be heavily geared towards numerical analyses and use vocabulary that is not necessarily familiar to humanities students. It can be difficult for humanities students in such classes to see applications to their research questions if their homework assignments task them with writing code to find prime numbers or enumerate the values in the Fibonacci sequence.

Overall, programming is a skill of its own developed over an entire career. As we teach young humanists to take advantage of digital data, we must not lose sight of the objective, which is for them to become better trained humanists, with access to more humanities data and the skills to explore that data. Additionally, their training should make them able to communicate more effectively with professional technologists, which is also valuable. However, it would not seem fair to ask our students to become expert programmers and expert humanists all at once, and we thought it important to stick to the priority in this introductory class, namely the humanities content of the class and its associated data.

With all this in mind, we laid out an alternative path where students could interact with data at a high level and produce in-depth analyses using a Graphical User Interface (GUI). This way, they could approach complex problems and learn high level techniques without

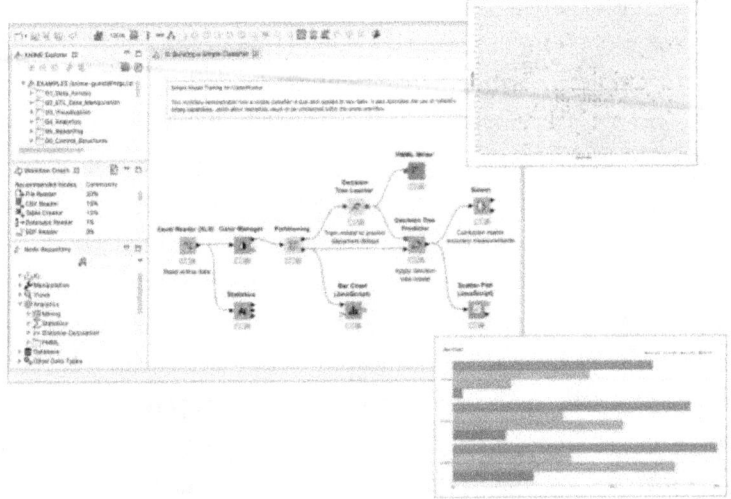

Figure 2. Screenshot of a KNIME Analytics Platform workflow illustrating reading data from Excel, manipulating it, and plotting it. From https://www.knime.com/knime-software/knime-analytics-platform

the added layer of learning a programming syntax. We chose the KNIME Analytics platform, a graphical workspace for building data analysis workflows. In their own words,

> KNIME Analytics Platform is the open source software for creating data science applications and services. Intuitive, open, and continuously integrating new developments, KNIME makes understanding data and designing data science workflows and reusable components accessible to everyone.[9]

In KNIME, users create *workflows*, which represent single units of data analysis (see Figure 2). Users drag nodes that read data, transform data, and visualize data into their workflow as appropriate and connect nodes together visually via drag-and-drop operations. At each step, users can configure any node to their own specifications (e.g. filter on a particular piece of text, take an average, etc). Users can also see the state of their data and analysis at it passes through their

---

[9] https://www.knime.com/knime-software/knime-analytics-platform.

workflow and troubleshoot as needed. The tool can perform almost any task a programming language can, but without the obstacle of learning an arcane syntax while simultaneously learning data analysis concepts, students can become fluent in KNIME within a few weeks.[10] The advantages for a class such as ours appeared obvious.

From the perspective of computer science education, KNIME neatly solves several issues and meshes well with emerging best practices. To the first point, it has been observed several times that among the concepts introduced in introductory computer science, functions and data types are among the most difficult.[11] In KNIME, a function is simply a node, and function application is a drag-and-drop connection between two nodes; thus, rather than being a brand new and puzzling concept, functions resemble actions students are already familiar with from using a graphical operating system like Windows or MacOS. Similarly, any piece of data in KNIME that has a type is clearly marked as such, again rendering an otherwise puzzling concept into a familiar form, tagging.

To the latter point, recent research suggests that *computational thinking* is a distinct skill from *coding*, and that the syntax of the latter is an obstacle to the development of the former.[12] KNIME allows us to focus on computational thinking in our course, conveying concepts like data, data types, transformations, and input/output separately from how those concepts manifest in a specific programming language. Students can leave our course with a firm grasp on computational thinking that will aid them in later Computer Science courses, should they opt to go that direction in their studies. For instance, students who later encounter the distinction between floating point numbers and strings—a distinction that is core to working with computers and will invariably come up in every programming language class—will already have experience working with this distinction. Students who later encounter databases and the SQL query

---

[10] This was initially a hypothesis, but it was later borne out by student outcomes.
[11] Hertz and Ford 2013.
[12] Parsons and Haden 2006.

language will have already encountered the notion of a table, sorting and projecting data, and grouping and aggregating, concepts that are challenging to grasp when they are first met

One final non-negligible advantage to this formula is that no coding skills are absolutely necessary on the part of the instructor. This means that in a situation where team-teaching were not possible, or if this instruction were to be integrated into another type of humanities class, such as a methods class, a humanities instructor without special training could conceivably conduct the class by themselves after a manageable amount of background preparation.

Our class

Three iterations of the class have run so far, in the fall 2016, 2017, and 2018. The class size was limited to twenty students due to the capacity of the computer lab. The student population included first-years who were part of the advising program at Tufts (Marie-Claire Beaulieu was their first-year advisor and got to advise them regularly as part of the class). Many upperclassmen, mostly in Classics, also took the class, as well as several graduate students in various humanities fields. Over the years, we have noted that all these constituencies benefited from the class, in particular, first-years, who tended to do particularly well and showed remarkable motivation and passion for their research topics. First-years regularly earned the top grades in the class.

The class was oriented towards research questions. From the first week in the semester, students were asked to form groups of two or three. Beaulieu conducted a visit of the Greek and Roman collections at the Boston Museum of Fine Arts, giving instruction on artifacts such as inscribed stones and vessels, ritual vases, and coins. These materials were examined in the context of large datasets such as LGPN, the Mantis numismatics database, and the CVA. For each, we asked whether they would appear in the data or not. For instance, we noted that CVA and LGPN include an entry for a fifth-century red-figure lekythos on display at the MFA showing Apollo and bearing the inscription ΗΙΠΠΟΝ ΚΑΛΟΣ ("Hippôn is handsome") (see Figures

Figure 3. Attic Lekythos, about 470–460 B.C., The Museum of Fine Arts, Boston, 95.45 (image: Public domain)

3, 4, and 5). The CVA includes the entry because it is a Greek vase, and LGPN because it bears an attestation of the name of a Greek individual. We therefore noted what these entries were and were not: CVA catalogs objects, and LGPN catalogs attestations of names, rather than actual *individuals*. Conceivably, LGPN could have included multiple attestations of the same Hippôn, if his name happened to be recorded on more than one preserved and catalogued object. In this way, students gained a good understanding of the nature of the data they were going to work with, and its limitations. We noted for instance that LGPN is subject to the biases of inscriptions: male upper-class citizens are more likely to be represented than females or lower-class individuals, given the expense associated with fine pottery or inscriptions. Similarly, CVA is strictly concerned with fine pottery and would likely not include ostraka or other lower-quality materials.

207426, ATHENIAN, Boston (MA), Museum of Fine Arts, 95.45

[View larger image] [Add to photograph album]

- **Vase Number:** 207426
- **Fabric:** ATHENIAN
- **Technique:** RED-FIGURE
- **Shape Name:** LEKYTHOS
- **Provenance:** BOEOTIA, THEBES
- **Date:** -475 to -425
- **Inscriptions:** HIPPON
Kalos/Kale: HIPPONKALOS
- **Attributed To:** PROVIDENCE P by BEAZLEY
- **Decoration:** BD: APOLLO WITH PHIALE AND KITHARA
- **Current Collection:** Boston (MA), Museum of Fine Arts: 95.45
- **Publication Record:** Beazley, J.D., Attic Red-Figure Vase-Painters, 2nd edition (Oxford, 1963): 1586
Beazley, J.D., Attic Red-Figure Vase-Painters, 2nd edition (Oxford, 1963): 640.74
Beazley, J.D., Attische Vasenmaler des rotfigurigen Stils (Tubingen, 1925): 135.45
Carpenter, T.H., with Mannack, T., and Mendonca, M., Beazley Addenda, 2nd edition (Oxford, 1989): 274
Maas, M. and Snyder, J.M, Stringed Instruments of Ancient Greece (New Haven, 1989): 78, FIG.19
- **AVI Web:** https://www.avi.unibas.ch/DB/searchform.html?ID=2791
- **AVI Record Number:** 2644
- **CAVI Collection:** Boston 95.45.
- **CAVI Lemma:** RF lekythos. From Thebes{1}. Providence Painter. Second quarter fifth. 470-460.
- **CAVI Subject:** Apollo with lyre and phiale, frontal and looking left.
- **CAVI Inscriptions:** To left of his face: κα[λ]ος. To left of his body, diagoally upward: hιπον καλος.
- **CAVI Footnotes:** {1} see Boston 95.43 and 95,44.
- **CAVI Number:** 2644
- **AVI Bibliography:** CB (1931–63), II, 42/88, pl. 46. — ARV[2] (1963), 640/74. — Add.[2] (1989), 274.

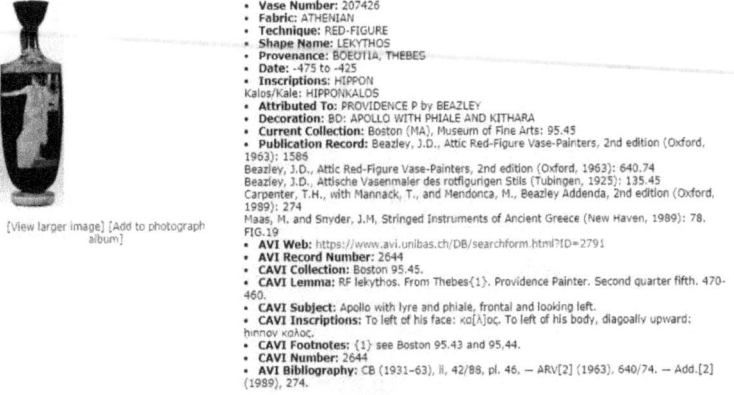

Figure 4. Corresponding entry in CVA for MFA lekythos 95.45

Download these results as: XML JSON KML CSV RDF

Table Map

Display 50 records Search:

ID	Vol.	PubID	Name	Sex	Place	Floruit	References
V2-35023	2	3	Ἵππων	[m.]	Athens	470-460BC	*ARV²* p. 1586 II (*hιππōν*: kalos)

Showing 1 to 1 of 1 entries

Figure 5. Corresponding entry in LGPN.

This deep dive into the relationship between data and real-world objects coalesced in a project to be completed over the whole term. After the museum visit, students were asked to choose an object or group of objects that they found interesting in the museum, and to start formulating a research question and hypothesis that they could explore with data-driven methods. The project was turned in to the instructors regularly for feedback throughout the term. The final installments required students to have produced a research question, hypothesis and thesis, a KNIME workflow manipulating the data in an appropriate way, a bibliography of scholarly books and articles related to the research question, and a detailed report explaining the findings and connecting them to current scholarship in the field.

In order to view the data in context, and to practice manipulating different types of data, the projects were required to include at least two of the datasets used for the class, namely LGPN, Mantis, the CVA, the Pleiades Gazetteer, and the Perseus Digital Library. An example of a particularly successful project was one asking, "How do the deities depicted on the coins struck by the adoptive emperors differ between mints at Rome and Alexandria and between different emperors?" by students Allyn Waller and Peter Spearman. The extensive workflow compared imagery found on the coinage of the adoptive emperors among themselves extracted from Mantis, along with, for Marcus Aurelius, a comparison of the imagery found on his coinage and the most frequent words in his *Meditations*, extracted from Perseus data. The workflow was accompanied by a thorough bibliography and a report detailing the findings.

In parallel to the term project, the class was organized in five modules distributed over a fifteen week semester which each focused on a technical skill and how to perform it in KNIME. Each module included three to four lecture periods, with two labs each with a lab report due at the end of each week, and two weekly quizzes focusing on class lectures and assigned readings. The modules were the following: 1) Tidy Data, 2) Visualization, 3) Probability and Statistics, 4) Natural Language Processing, 5) Machine Learning, Classification, and Prediction. Each of these technical skills was paired with a topic of inquiry in Classics. Bucci would demonstrate a particular technique in KNIME, and Beaulieu would lecture on the substance of a particular dataset, relating it to the ancient world, and demonstrate a research question that could be pursued in KNIME using the data and technique in question.

For instance, during the data tidying and cleanup module data cleanup was demonstrated using the LGPN dataset, where the "floruit" column, while mostly readable to the (practiced) human eye, is impossible to process by computer. The codes to express the active dates of the individual recorded in each attestation include both exact dates and ranges in Arabic numerals followed by letters (e.g. 185 AD, 119/18 BC, 119-18 BC), centuries in Roman numerals (e.g. ii BC, ii-iii AD, ii/iii AD) and statements such as "192 or 191 BC". An attempt was made by the editors of LGPN to counteract this issue by including two columns named "not before" and "not after", encoded in Roman numerals and using a minus sign to indicate dates before the common era. However, careful observation revealed inconsistencies and imprecision. For instance, "231 or 206 BC" is recorded as -206 in "not before" and "not after". In addition, arbitrary choices were made to represent period statements such as "a. 133BC", which was taken to mean a range of +50 years, and therefore recorded as not before 133 BCE and not after 83 BCE. Marie-Claire Beaulieu designed an extensive but simple workflow (Figure 6) designed to separate each case and normalize it into to four columns, some of which would or would not have a value. A decision to create missing values was thus deliberately made, the pros and cons of which were discussed in class. The students expressed interest in data cleanup and reported the usefulness of a thorough overview of the various techniques available, as data cleanup famously takes up 80% of a data analysis project.[13]

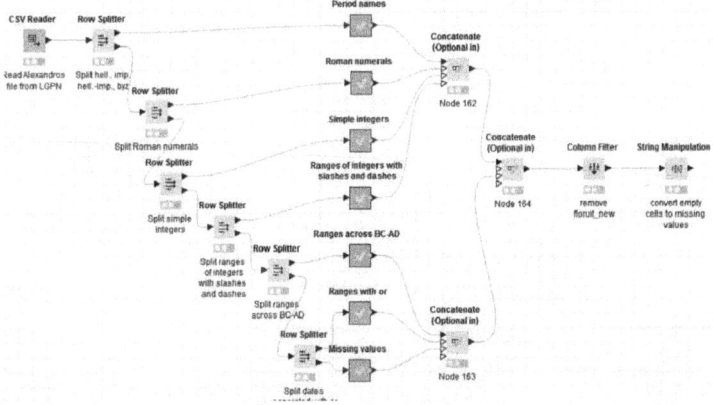

Figure 6. KNIME workflow to normalize a subset of the LGPN data (the 2551 entries for attestations of the name Alexandros). Each gray metanode contains a subordinate workflow which manipulates the data to normalize it. Typical operations include string replacement, cell splitting, type conversions, etc.

This organization of the materials, with one teacher focusing on techniques and the other on humanities content and examples, allowed us to take full advantage of team-teaching. Each class was taught with both instructors present. One instructor would lecture and the other would participate in the discussion by asking questions or making connections with previous materials. In this way,

---

[13] Dasu and Johnson 2003; see also Wickham 2014.

we modeled the interaction between humanists and technologists as they work together to solve a problem, a setup that our students are likely to encounter later in their careers as they join diverse work teams.

Sample module: machine learning and classification

One of our most appreciated modules was machine learning and classification, as it addressed a topic that often makes the popular news and carries many misconceptions about its capacity for fair, impartial judgment or the possibility that machines will take over the world in our lifetimes. The students enjoyed learning about this, and also performing a machine learning task themselves with KNIME.

In order to demonstrate the application of machine learning to a humanities problem, we chose to discuss the placement of religious sites in the landscape of the Greek world, especially with respect to *poleis*. This fascinating question has been much debated in classical scholarship. François de Polignac's book *Territory, and the Origins of the Greek City-State* (1995) was used as a guide to explore the question from a scholarly point of view. Polignac investigates the origins of cults set up in outlying territories and the role they played in the development of urban centers. He challenges the commonly held view, inherited from philosophers such as Aristotle, that cities were formed from the disintegration of clan/monarchic systems into a polis system and that the establishment of the poliadic divinity at the city center subordinated the rest of the community's cults. Rather, Polignac emphasizes the changes in religious life that occurred at the end of the 8th century BCE, marked by new types of offerings and multiplied cult sites with monumental buildings. He notes that a large number of these are outside the city and concern the cult of heroes, or rites for the integration of the youth, foreigners, and women. He concludes that these changes were connected to the development of non-urban cults and that the polis was established through the religious definition of a new space, namely the city territory (*chora*). A new civic community was articulated through the mediation of religion, in particular at the edges of the territory, via social integration rites. The book is an important and much debated milestone in the field, and also has the merit of being short and relatively accessible for students who are not in classics. The method Polignac employs is also interesting: he proceeds by case study, focusing in particular on the Argive Heraion, partly in an attempt to move away from the exceptional Athenian experience.

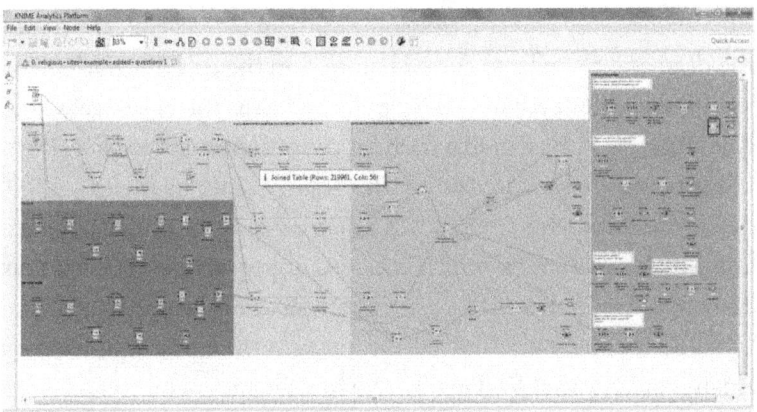

Figure 7. General outline of Sanctuaries vs. Cities workflow.

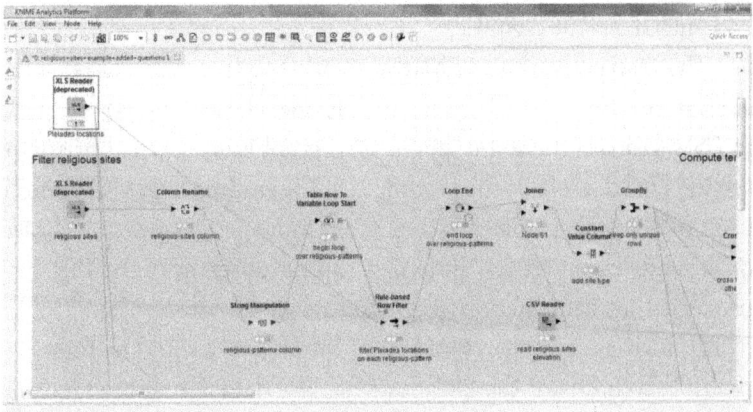

Figure 8. Extracting religious sites from Pleiades data.

Our goal was not to replicate Polignac's argument exactly or test his conclusions, but rather, to see if we would get different results using a digital dataset and a classification algorithm, and most importantly, to explain the reasons behind these differences. As preparation for the in-class demonstration, students were asked to read the first chapter of Polignac's book and to discuss it in class. They were also quizzed on the general argument as part of the weekly quiz. Beaulieu lectured on the Greek religious system, insisting in particular on the different types of cult and cult sites. For the purposes of the demonstration, she differentiated between three broad groups of sites, namely large sanctuaries, some of them panhellenic, the cult of the dead and the related cult of heroes, and the cult of agrarian and nature divinities such as the Nymphs and Pan. The objective was to give students a broad preparation for the question and to allow them to form hypotheses based on the information they had.

In class, students were shown a map of some of the most important Greek sanctuaries, which included topographical information. They noted an imperfect but significant relationship between elevation and the deity worshipped, for instance the worship of Zeus on Anatolian Ida and at Dodona. Thus, a further parameter was added in the analysis of the sanctuary data, namely whether elevation played a role in the placement of sanctuaries and what connection elevation had with different divinities.

The instructors built a KNIME workflow that extracts religious sites, ports, and urban areas from a Pleiades data dump downloaded in CSV format. These are represented graphically as the horizontal sections on the left of the workflow in the outline below (Figure 7). Each of these sections extracted sites from the Pleiades data based on a predetermined list. Indeed, a careful observation of the data revealed that information about the religious nature of a site (e.g. words such as "temple", "sanctuary", "tomb", etc.) could be found in multiple columns of the Pleiades data. The data is first read in with a Reader node and then filtered in a loop construct against a list of such words (see Figure 8). An additional CSV reader node adds elevation data (obtained externally) and joins it to the sites after the sites have been filtered from the raw data. The same process is repeated for the ports and for urban areas. Within these sections, we were able to show students the concepts behind a loop structure, a powerful tool in computing, whatever language may be used. We also demonstrated the joining of two datasets, which allows us to find answers to our specific question.

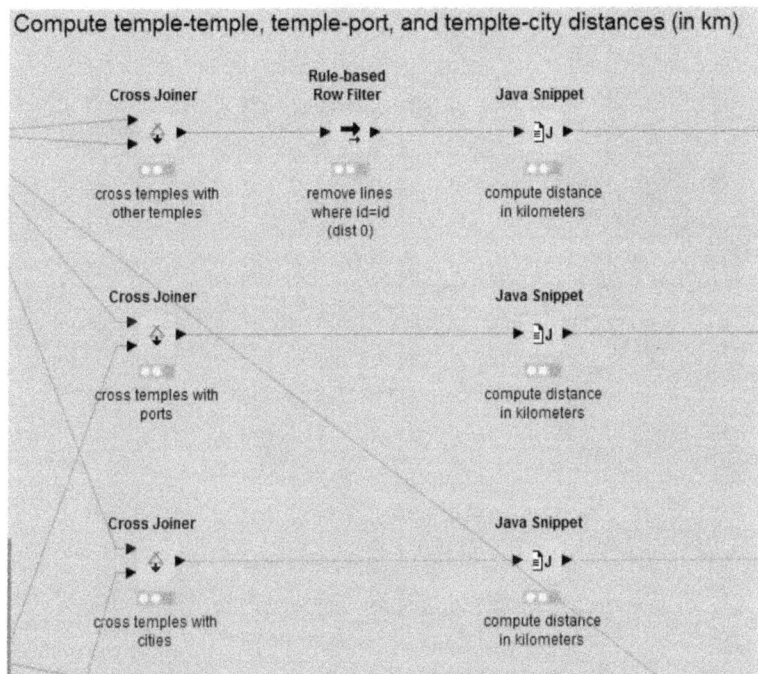

Figure 9. Compute distances between religious sites, cities, and ports.

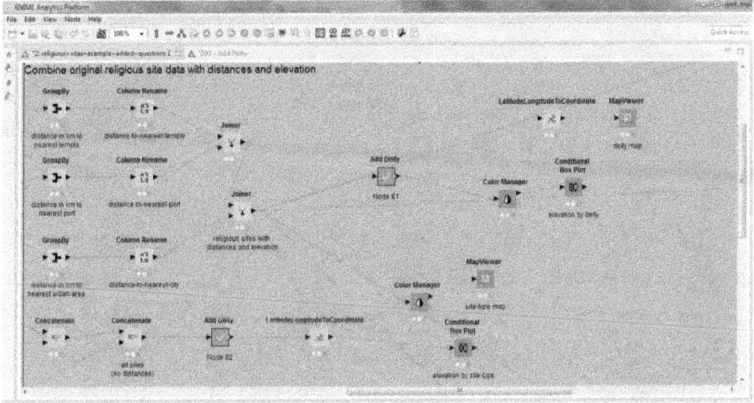

Figure 10. Combine distance data with elevation to provide visualizations addressing the research questions.

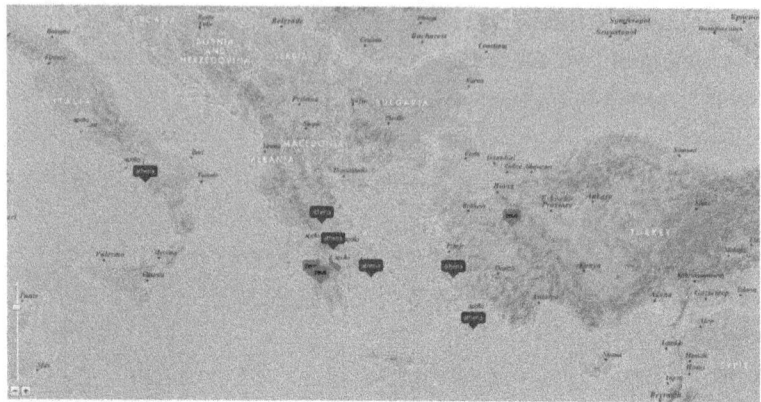

Figure 11. Map showing the locations of the religious sites associated with Zeus, Athena, and Apollo found in the Pleiades data.

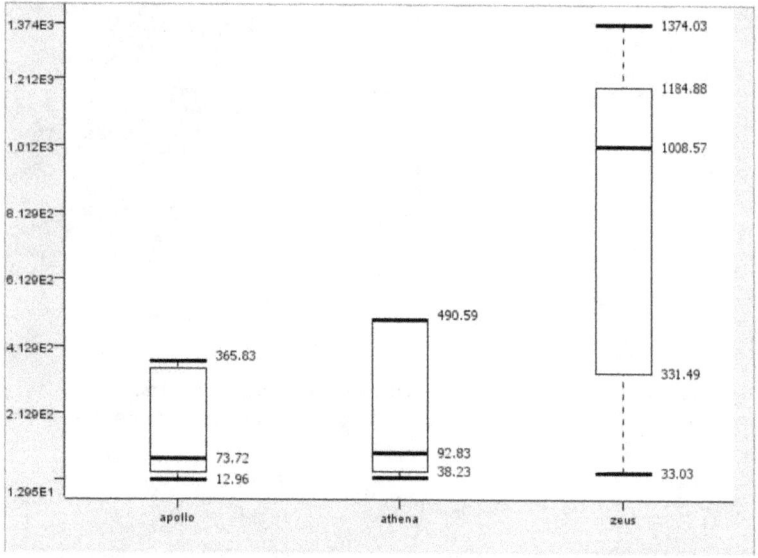

Figure 12. Elevation of religious sites connected to Zeus, Athena, and Apollo

The middle section of the workflow (see Figure 9), computes distances between religious sites, ports, and urban areas. Each type of site is cross-joined with itself (making sure to remove identical locations) and then with cities and ports to have a full dataset of distances.

Finally, the last part of the workflow (see Figure 10) combines the data manipulated and obtained in the earlier steps and provides answers to the research questions via two visualization types, namely a map (Figure 11) and a box plot (Figure 12). The conclusions found this way confirmed the students' intuitions concerning Apollo, whose temples were noted to often be near the coast. Similarly, temples of Zeus seemed to be, in general, located at much higher elevations than those of the two other deities studied here.

These initial findings were encouraging, and they offered a way to confirm hypotheses with data-driven answers. They also offered a means to loop back to Polignac's work and assess the differences between his findings and ours. In particular, we wondered whether data issues could be obscuring our findings. In a set of additional manipulations, we looked at the distribution of religious sites by period, to address Polignac's claim that great changes occurred in the Archaic period. A large number of religious sites indeed surround cities at that period, compared for instance with the Hellenistic period, during which it appears that a large number of sites connected with the cult of the dead surround urban areas. However, in examining this finding, we noted that we computed the Pleiades data with the earliest known date of existence of a religious site, rather than its whole span of existence, and we may thus be missing crucial information. Additionally, qualitative information is absent from this raw analysis, such as the waxing and waning of the levels of frequentation or influence of any particular religious site throughout time. These observations are interesting pedagogically, as they demonstrate the need to be critical of one's findings as well as the need to work in iterations from initial findings to refined conclusions.

Lessons learned

Over the three successive iterations of this class (fall 2016, 2017, and 2018), we changed our approach in several ways. After the first round, we introduced weekly quizzes, as we noted that students were not always doing the required readings or fully grasping important concepts seen in class. From the humanities standpoint, we also noted that a general introduction to the ancient world or readings from textbooks such as *Ancient Greece: A Political, Social, and Cultural History* by Pomeroy et al. (2011), were too broad to offer students who were not in classics an opportunity to take hold of a topic. We instead started focusing on smaller questions, such as specific authors or works, or types of materials. For instance, during the module on Natural Language Processing, we conducted a comparative study of the writing styles of Plato and Xenophon, and introduced both authors without going into an introduction to ancient philosophy. Similarly, in the statistics module, we examined datasets of ancient vases and gave an introduction to the forms and functions of Greek ritual vessels, without going into great detail about the rituals themselves.

After the first year giving the course we felt the learning curve for KNIME was still a bit too steep. We introduced a homework assignment inspired by Parson's puzzles[14] in which we placed the appropriate KNIME nodes into a workspace and asked the students to draw the required connections among the nodes. To complete the assignment, students would need to understand the reasoning behind the workflow, but since we placed the nodes for them they were not required to generate the entire logical flow themselves. We also felt that students groups had a tendency to adopt a division of labor, in which one team member became the "workflow expert" while the other team members contributed little if anything to the final workflow. In the third year we made mastery of the workflow a soft requirement for all team members, meaning we did not grade for it but we did check in several times during the semester and verified all team members understood their team's workflow.

Class outcomes

We are very proud of the outcomes of this class for our students. While not every student took advantage of this introductory class with the same dedication or degree of success, those who did saw a definite turn in their studies and future careers. Several of our first year students joined Tufts Enigma,[15] an independent data analysis journal on campus, where they were able to hone their skills in producing complete data analysis papers using a variety of tools. Several more found the class helpful in deciding to pursue a degree in Computer Science, whereas before they had been interested but somewhat intimidated

---

[14] Parsons and Haden 2006.
[15] http://tuftsenigma.org/.

at the prospect. Finally, one graduating senior in Classics obtained a job as Data Quality Coordinator at a major international charity due to this experience. Another Classics senior felt empowered to pursue in-depth analysis of political data, which she had been interested in before but did not have the tools to conduct. She continued on this path after the class ended, eventually learning R through R Studio, and finally landing a job as Data Analyst at a major national non-profit. We find these two latter outcomes particularly encouraging, as they show that even basic skills in data science, when paired with a humanities education, reinforce each other and offer career outlets which give the humanities pride of place. In today's world, especially when it comes to curbing the excesses of culture and ethics-agnostic technology, and also filling positions which require a high level of understanding of society, politics, and art, a strong influx of humanities-minded individuals is desperately needed in the workforce.[16]

From a pedagogical standpoint, setting this class up and teaching it was a challenge which we are happy to have tackled. Certainly multiple other approaches could have been taken, but we feel that we were successful in truly melding an introductory Computer Science class with a Classics class, where students learned at the intersection of both disciplines. We found that a great deal of humility and flexibility is required to take real advantage of a team teaching scenario, where both instructors teach each lesson. We also found that, perhaps because the class tends to sow seeds rather than provide an immense influx of new knowledge, the outcomes for students only become fully realized in time, typically a year or two after the end of the class.

Conclusion

The decision to not teach programming was beneficial, in our view, but not all students agreed, reflecting the current debate on this question in the field of Digital Humanities. There may also be a gender aspect to the debate. Male students tended to drop the class more

---

[16] See https://mellon.org/resources/shared-experiences-blog/good-job-humanists/ and https://culanth.org/articles/966-what-should-an-anthropology-of-algorithms-do.

than female students, and anecdotally, many of those who dropped told us they were looking for a programming class. On the other hand, many female students excelled in the class, particularly in the technical tasks, which challenges commonly held gender notions regarding technology.[17] We argue that this does not represent the fact that the women did better with the "soft" option of a GUI vs programming, but rather that they focused on computational thinking rather than get bogged down in programming syntax. Additionally, as beginner technologists, they may have been more realistic about the learning process and the steps to follow to achieve their goals, unlike some of their fellow male students who chose to dive into programming with limited preparation or thought they could acquire a programming language with just one class, and struggled as a result. In a broader perspective, our students' successes also challenge the commonly-held belief that humanities and technology are mutually exclusive, which unfortunately leads many humanities students to self-select out of technical fields. We hope to help reverse this trend, and also to send an influx of humanists into technical careers, where their training in ethics, culture, history, politics, and arts is greatly needed to complement the skills of professional technologists and form well-rounded teams.

Works Cited

Dasu, T. and T. Johnson
2003 *Exploratory data mining and data cleaning*. Wiley-IEEE, New York.

Hall, Macie
2015 Bringing Digital Humanities into the Classroom. *The Innovative Instructor*, April 3, http://ii.library.jhu.edu/tag/digital-humanities/.

Hertz, Matthew, and Sarah Michele Ford
2013 Investigating factors of student learning in introductory courses. Proceeding of the 44th ACM technical symposium on Computer science education. ACM, New York.

Kingsley, Jennifer
2016 A Low Tech Approach to Digital Literacy. *The Innovative Instructor*, July 2016, http://cer.jhu.edu/files/InnovInstruct-Ped_approach-to-digital-literacy.pdf.

Leigh Bonds, E.
2014 Listening in on the Conversations: An Overview of Digital Humanities Pedagogy. Special Issue, *The CEA Critic*, 76:2.

Liberal Arts Computer Science Consortium.
2007 A 2007 model curriculum for a liberal arts degree in computer science. *J. Educ. Resour. Comput.* 7(2): Article 2.

Mahony, Simon and Pierazzo, Elena
2012 Teaching Skills or Teaching Methodology? In *Digital Humanities Pedagogy: Practices, Principles and Politics* edited by Brett D. Hirsch, , Open Book Publishers, Cambridge, https://www.openbookpublishers.com/htmlreader/DHP/chap08.html#ch08.

Parsons, Dale and Patricia Haden
2006 Parson's programming puzzles: a fun and effective learning tool for first programming courses. In *Proceedings of the 8th Australasian Conference on Computing Education - Volume 52* (ACE '06), edited by Denise Tolhurst and Samuel Mann. Vol. 52. 157-163. Darlinghurst, Australia.

Polignac, François de
   1995 *Cults, Territory, and the Origins of the Greek City-State*. Translated by Janet Lloyd. University of Chicago Press, Chicago.

Ramsey, Stephen
   2012 Programming with Humanists: Reflections on Raising an Army of Hacker-Scholars in the Digital Humanities. In *Digital Humanities Pedagogy: Practices, Principles and Politics* edited by Brett D. Hirsch, , Open Book Publishers, Cambridge, https://www.openbookpublishers.com/htmlreader/DHP/chap09.html#ch09.

Wickham, Hadley
   2014 Tidy data. *The Journal of Statistical Software* 59: https://www.jstatsoft.org/article/view/v059i10

Digital Creation and Expression in the Context of Teaching Roman Art and Archaeology

Sebastian Heath

My purpose in this paper is to present the digitally-informed pedagogic goals and some of the specific digital practice that I have integrated into my teaching of Roman Art and Archaeology courses at the undergraduate level. In addition to general courses on Roman archaeology, I draw on my experience teaching *Rome: A Visual and Virtual Empire* in New York University's Gallatin School of Individualized Study and then *Digital Approaches to Roman Art and Archaeology* in the Art History Department at NYU. I should stress, however, that I am not describing the syllabus of any one course, nor even faithfully hewing to individual assignments that I have set for my students. The Gallatin course was taught in Fall Semester 2016 and the Art History course in Fall Semester 2018. Because my primary appointment at NYU is at the Institute for the Study of the Ancient World, I was a visitor in both departments and I am grateful for the opportunity each provided me to engage with undergraduates. The particular tools I used have all changed to at least some extent since the teaching I describe here took place; indeed, a few are no longer available. New tools have also come into existence. My response to this dynamic situation is to start with general principles, to highlight content in the expectation that it will remain relevant, and to describe the tasks I gave students in such a way that their underlying purpose is evident so that changing tools can be adopted in future classroom settings. It is the case that both courses that I draw upon here were taught in digital labs that provided a computer for each student taking the course. *Rome: A Visual and Virtual Empire*, taught in 2017, had 20 enrolled students, which is the maximum I would be comfortable teaching without assistance. But I don't believe that every task and assignment I describe requires that every student have in-class access to a machine that they can use on their own. Nonetheless, that certainly is a relevant circumstance that readers should be aware of as they consider what follows.

An opening general statement of digital teaching philosophy is as follows: I look for opportunities to combine the processes by which students learn a digital tool or method with learning about the material culture of the Roman empire in its political and cultural context. In short, I try to combine "doing" and "learning". To some extent I adopt the overlapping pedagogic practices known as "project-based learning" and "experiential learning", though I do not mean what follows to be a focused discussion of how to apply any particular approach in class. I will be anecdotal in tone, even though my "anecdotes" are adapted from my experience and are not, as indicated above, a direct reporting thereof. In this paper I will focus on 3D modeling and also discuss mapping—along with the overlap between the two—as digital approaches that allow me to put this philosophy into practice.

Using 3D to Create and Learn

As is likely familiar to readers, 3D representations of physical objects and built spaces are compelling, and this very general observation applies to Roman material culture of the imperial period as much as it does to any other field of study.[1] The specific digital method that I lead up to here is student use of photogrammetry to make 3D models from sets of photographs that they have taken. Along the way, I will describe how learning 3D skills can be integrated into teaching in a way that allows students to engage closely with imperial portraiture and with related topics such as how skin tone was depicted in ancient art and how that affects student restorations of ancient polychromy. These are not uncomplicated issues to address and this paper is not intended to lay out a single script that in-class discussion can follow or to restrict what ideas students should grapple with in completing

---
[1] Bond 2017b.

their own assignments. My audience is instructors and my goal is to describe a digital method for bringing these current issues into digital coursework.

When putting teaching philosophy into practice I introduce 3D content and skills so that in-class work allows students to move along a trajectory of 1) acquiring 3D content, understanding what it is and its limits, and how it supports discussion of the object represented; 2) editing that content to allow students to communicate their engagement with relevant readings and to show their understanding of the underlying object; 3) creation of new 3D content from scratch, with this activity again providing an opportunity to learn about the

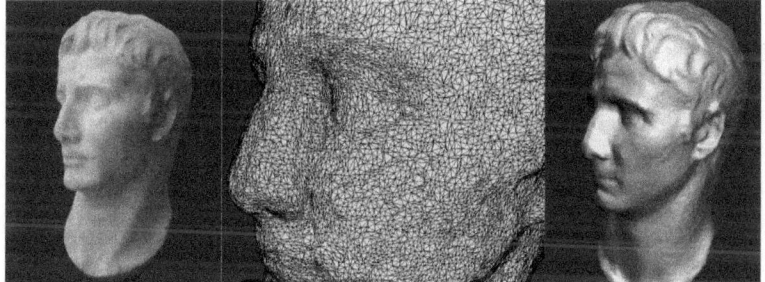

Figure 1: Renderings of the Getty Augustus portrait (Acc. no. 78.AA.261). From right to left: (a) Fully rendered, (b) Wireframe, and (c) "matcap."

objects they are modeling. While I have introduced this sequence in the context of 3D data, I mean it to be a general rubric for introducing digital content or digital method.

In 2019, as I write this paper, the availability of 3D content is an established but still relatively new development. Currently, the commercial site SketchFab is an easily accessible source of relevant models, and the site also provides an expanding set of tools for displaying and interacting with 3D content. As a side note, I do think that academic users should be wary of excessive dependence on a single provider whose tools are not open-source, but I put that particular concern aside for now as I focus on using SketchFab in the classroom.

It is typical for me to start my discussion of 3D content by asking students to view the model of the Roman Emperor Augustus that is shown using various renderings in Figure 1. Indeed, showing those different views of the object is part of the class discussion. Fig 1a (at left) is the default view. It is "fully rendered", indicating both the form and surface color. A first step is for students merely to open this view on their screens. I note that there are three basic ways of interacting with 3D models that they can expect any 3D-aware software to support. Those are: rotation, zooming, and panning. I do stress that the specific keyboard combinations that initiate these interactions can be different in any one tool, which can be a little frustrating. My point here isn't to be bogged down by detail, but instead to indicate that the collective ecosystem of digital tools is variable and imperfect. Coming to understand this general circumstance, gaining the confidence to work around it, and creating a classroom environment in which students help each other are processes I want to begin early in the course.[2]

The other renderings in Figure 1 (b and c) support in-class discussion of "What is a 3D Model?", at least from the technical perspective. 1b is the "wireframe" view zoomed in to show that underlying many models are the concepts of "vertices," or xyz points in a virtual 3D dimensional space; these vertices are connected by "edges;" 1b shows the edges forming triangles that are known as "faces." Faces in turn can be colored in response to how an artificial light hits them, which is the rendering shown in 1c, also known in SketchFab and other tools as the "Matcap" view. Returning to Figure 1a, the default view, shows that color is exceedingly important in allowing a human viewer to perceive digital content as a surrogate for a physical object.[3]

In these early sessions with 3D content my goal is to communicate to students that there is no "magic" involved in presenting compelling representations of Roman material culture such as the portrait of Augustus now in the Getty (Acc. no. 78.AA.261). I have found that most undergraduate students I teach have not gone beyond using computers for writing papers and browsing the web; though they certainly have embraced all the richness of media that browsing the web can bring to their screens. That familiarity with digital content does put students in good position to integrate a small amount of technical terminology into their understanding of the digital resources I am

---

[2] Watrall 2019
[3] Brooks 2017

asking them to use. Small and tractable conceptual steps give students a workable sense of what is going on "under the hood." In this paper, I mean to highlight that finding opportunities to demystify digital data is another useful pedagogic approach. A result of showing the different renderings in Figure 1 is in-class discussion of the limits and potential of digital surrogates of Roman material culture.[4] How close can we zoom in on details of Augustus' eyes or on the curls of his hair? We can examine the smoothness of his skin and think about the

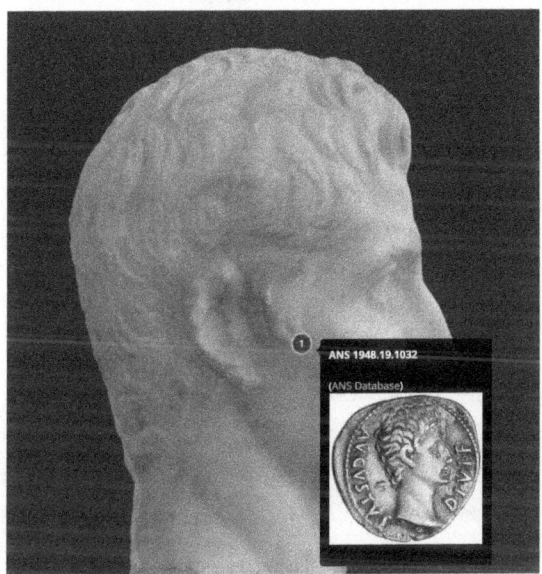

Figure 2: Detail of model of Getty Augustus in Sketchfab with annotation showing. The coin is in the collection of the American Numismatic Society (1948.19.1032).

amount of work it took to create the portrait. These are details that support initial thinking about intent at the time of creation and the experience of subsequent viewers.

Having established some understanding of the technical aspects of 3D content, I move quickly in class to re-using it. SketchFab provides an easy step beyond viewing models to support this process. Specifically, students can make their own free accounts, download models that are set to allow such access - including my model of the

---

[4] Gartski 2017, Rabinowitz 2015

Getty Augustus - and then re-upload that model to their own account. The specific reason to ask students to download and then re-upload models is to put them in position to use SketchFab's "annotation" functionality. Figure 2 illustrates a simple example of annotating a model. It shows a pop-up window that appears after a user clicks on the small circle that is labeled "1". That pop-up in turn shows a coin of Augustus and a link to further information about that object in the collection of the American Numismatic Society. I do not illustrate the SketchFab interface for making annotations here as it is likely to change, but it is not overly complicated. It is useful that not only can users place annotations on the virtual surface of a model and edit the content of the associated pop-up, they can also set the camera angle and zoom from which the model is viewed when the circular annotation icon is clicked. This is a small but important capability and introducing it to students sets up the first assignment using 3D data that I ask them to complete.

That assignment is actually quite simple. In conjunction with a reading, I ask students to choose a model to download from SketchFab, to upload it to their account, and to add meaningful annotations that show their understanding of the article I assign. Diana Kleiner's *Semblance and Storytelling in Augustan Rome* raises relevant issues and allows students to supplement its limited illustrations with their own digital work. In particular, Kleiner's discussion of hair-styles and the role they played in dynastic representation can be illustrated by annotating models and setting views that illustrate the complexity of both male and female coiffure. As for content at the time of this writing, the British Museum provides a good selection of downloadable models. Livia's appearance is a focus of Kleiner's article and a model of the BM's portrait of her is available for download (https://skfb.ly/6OpW9). Many students will create an annotation that quotes Kleiner and zooms right into a relevant detail of that portrait. I find it compelling that it is a virtual representation of an object that allows students to engage so closely with its underlying materiality and I am not shy about sharing my excitement in class.

Linking to Wikipedia from SketchFab annotations is also common. I encourage this. One requirement of the assignment is to post a link to their work into a shared Google Doc. At the start of class by which students have had to complete this work, we look at the annotated models together. One goal of this discussion is to talk about which sites on the public Internet make for good end points of links. As part of setting up this assignment, we will have had a discussion that includes my communicating that Wikipedia is acceptable, though students do need to make a determination as to the overall quality of any given Wikipedia article. Sites with too many advertisements are bad, and ones that will show a pop-up on my computer when I first follow the link are forbidden. Newspaper articles can be useful for illustration and for their reporting of recent discoveries, but students do need to be wary of repeating any broad conclusions about the Roman Empire that appear in them. My goal is not to set hard-and-fast rules. Instead, I want to allow my students to become thoughtful evaluators of the quality of internet sites. They have agency as they use digital content to demonstrate their understanding of a reading. As I move forward in this paper, "agency" will remain an important theme. By making their own specific choices within a digital ecosystem that allows them to work with the combination of content and tools, students communicate to me that they have thoughtfully engaged with the assignments I set for them. I am looking to find their thought and their care in whatever form it takes.

Having built familiarity with downloading, sharing, and annotating 3D models with links to other content, I then introduce students to more direct manipulation of these models and to more intimate choices about the ancient world. Excerpted from the context of the study of Roman society, the digital skill I am teaching is how to virtually "paint" 3D models. The aspect of the Roman world with which I am asking them to engage is the diversity of skin tones and the ways an ancient viewer might have perceived that variation. An enabler of this section of the course is the ready availability of relevant secondary bibliography, open-source visual resources that support broad engagement with the topic, and an open source digital tool that allows students to express on screen the choices they make.

Figure 3: (A) Romano-Egyptian Mummy portrait (Getty 74.AP.11); (B) Portrait of couple from House of Terentius Neo in Pompeii, Wikimedia Commons; (C) Wedding scene from House VI.17.42, Wikimedia Commons ; (D) Glass vessel with scene of date harvest from Bagram, Afghanistan; Wikimedia Commons.

The secondary literature I have used starts with Sarah Bond's freely-available essay *Whitewashing Ancient Statues: Whiteness, Racism And Color In The Ancient World*, which remains an accessible introduction to the topic of polychromy on ancient sculpture and its relevance to ongoing debates in contemporary society.[5] It is especially useful that her article includes a large detail of a Roman-period Egyptian funerary portrait, reproduced under the terms of the Getty's Open Content program, as students do respond well to these intensely personal images of ancient people. Wikimedia commons is a source for

---

[5] Bond 2017a

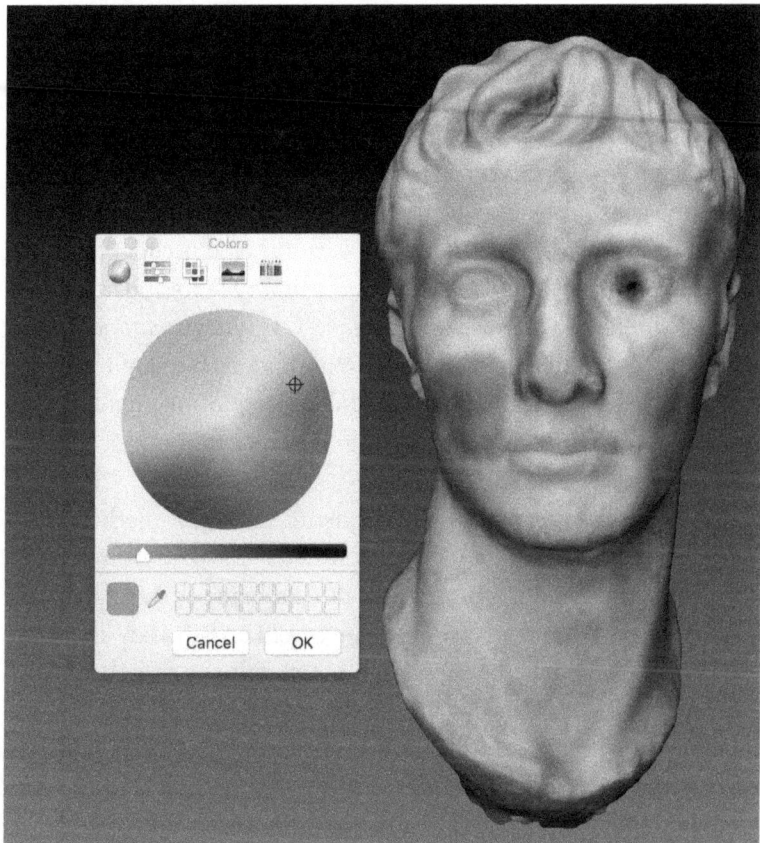

Figure 4: Trying different colors to paint the model of the Getty Augustus in Meshlab.

additional imagery, with digital versions of Pompeian wall-paintings being particularly useful. Figure 3 juxtaposes four examples of richly colored Roman-period representations of human figures that collectively hint at the variation in skin tone that existed in the ancient world and also hint at the in-class conversations to which including this content can give rise. Figure 3a is the portrait also included S. Bond's *Whiteness* article; the Getty has made this image available for reuse without restrction. Figure 3b shows the famous couple from the House of Terentius Neo at Pompeii. Figure 3c is another Pompeian wall-painting, this one from the so-called "House of the Golden Bracelet" (VI.17.42) and depicting a scene identified as either the Marriage of Mars and Venus or of Alexander and either Stateira or

Roxanne. Figure 3D shows a Roman-period glass vessel found at Bagram in Afghanistan, though probably made at an Eastern Mediterranean workshop, with a scene of date harvesting. The latter three images are all available on the Wikimedia Commons. Taken together, these images support in-class discussion of the representation of skin tone as it provides evidence for the ancient perception of such descriptive categories as gender, heroic status, and occupation in addition to ethnicity. There is a rich set of primary texts and secondary readings for all these issues. For example, Ovid *Ars Armatoria* 1.723ff highlights different expectations for male and female skin tone with reference to sailors and lovers. G. Woolf's recent *Strangers in the City* is a accessible contribution on ancient perception of ethnic identity and otherness.[6]

With this content available to students, the digital method I next introduce is the painting of 3D portraits downloaded from SketchFab. There a number of tools that can do this. In my classes, I have had success with the open-source software MeshLab, which is known as the "Swiss-army knife" of 3D tools. It is certainly the case that MeshLab has an idiosyncratic interface and limited documentation. I value it because it is an important part of the 3D open-source software ecosystem and I am enabling students to be effective users of 3D data by giving them experience with the tool. This is an outcome that is independent of any learning about the ancient Mediterranean world. Figure 4 cuts right to the chase of how working with MeshLab—or any other software that is able to 'paint' 3D models—contributes to students engaging with images and articles similar to those discussed above. It represents an early stage in showing students how to work with the software and is similar to work we do together in class. Specifically, it shows the Getty Augustus portrait after I have added various patches of color and worked on the detail of the emperor's eyes. I have purposefully made sure the "color-picker" is shown because that is key to the exercise. When students choose a color from that tool with the intent of applying it as the skin tone of a portrait, they are making a specific interpretation about the ancient world. Indeed, just the decision to add color to a stone surface is an

---

[6] Woolf 2018

interpretation. This last point follows from the circumstance that while scholarly consensus means it is no longer tenable to think of an ancient world filled with unpainted marble, it is certainly not the case that every sculpture was painted, and not the case that every painted surface perfectly retained that color during the years - even centuries - when it was displayed. It is the tension between the ancient evidence giving us options and the students having to make a choice that can generate good in-class conversation and lead to visually compelling student work.

I can briefly summarize the assignment that the students are asked to complete after I have shown them how colorize a 3D model: 1) download a 3D model of a portrait, 2) add color in MeshLab, 3) upload the painted model to SketchFab, 4) add annotations as relevant, 5) write a paragraph describing your work, including discussion of why you chose to color the portrait as you did. That description is very general. I do provide a list of downloadable 3D portraits that have overlap with the readings to date. This point leads to an important consideration: the range of work that the students produce will likely be influenced by when in the semester this assignment is given. In my teaching I have moved quickly towards "virtual painting" as a way to introduce the point that any act of digital creation implies interpretation and requires active choice. Setting this work early means that most students have colorized Julio-Claudian portraits, including the BM Livia portrait mentioned above. Delaying this assignment or re-arranging the order in which content is introduced could extend the corpus of 3D models to include portraits of Severan family members and the attendant expansion in the ethnic identities that students can consider.

I should note that there is no single correct digital response to even the small collection of visual and written material that I have illustrated and cited here. Students hand in diverse work. There is, however, an opportunity to allow digital tools to be part of the process by which students learn and think about the topic. I strive to be fair in recognizing student effort when I am grading their work. And, by being open ended in my description of how I go about creating that opportunity, I hope here to help others think about how similar assignments can be integrated into their own courses.

The next stage in my series of assignments based on 3D content is to ask the student to create a model on their own. This brings the sequence of "acquire, edit/use, and create" to an intermediate conclusion that I try to reach near the mid-point of the semester. The specific method I introduce is known as "photogrammetry" or "structure from motion" (SfM) and involves deriving a 3D model from a set of photographs of an object or physical space.[7] There are many introductions to photogrammetry so I will only emphasize here that undergraduate students can learn the technique and make a model that, while not perfect, is a solid basis for both exercising their 3D skills and for exploring an object. It is important that by the time I do introduce photogrammetry, students are very comfortable with 3D content. Again, at the very least, they can download it, make annotations, digitally paint, and even apply other basic edits such as resizing, rotating, duplicating whole models, and cutting out sections of a model that are not needed for their purpose. In this paper, I have been direct about how I have taught some of these skills in class. I do introduce students to the others and note here that the free software MeshMixer is a useful tool, particularly for duplicating and editing models.

I devote a sequence of two class meetings to teaching photogrammetry. To begin with, I usually return to the Getty Augustus portrait and show the set of photographs I used to make it. As recently as Fall 2018, Agisoft Photoscan is the software that I used. That application has now become Agisoft Metashape and the licensing terms have become more restrictive so that it is likely that the next time I teach this technique I will adopt an open-source alternative as these are becoming more user friendly. Regardless of the specific software, my goal is to demonstrate to students the making of a 3D model from photographs. Using "low quality" settings this takes only a few minutes. With the basic principle in place, I ask students to download a set of photographs that I have prepared and that I know will lead to a reasonably good model. Indeed, I have put four such sets online—all of objects and buildings in or near Rome—for anyone to use (the links are listed below). Once students have seen how the process works and

---

[7] Olson et al. 2013

have tried it in class, I give them the assignment of making a first model for the next class. I stress that it does not matter what it is of. We have, of course, discussed what kinds of objects or spaces work well for photogrammetry ("nothing too shiny and no glass"), but my goal is for the students to get any practice with the technique. They do have to upload their work to SketchFab, but that is a basic skill with which they are familiar. In the next class, we look at their models together and discuss the issues and problems they encountered. And we also enjoy their successes. This stage need not take a full class session.

All this work is preliminary to the students being assigned the task of going into a museum gallery in New York City and making models of two Roman objects of their own choosing. I ask that those models be thematically related in some way, though, again, I give students the option to set their own pairing. Obviously, that New York is home to two major collections of ancient art—the Metropolitan Museum and the Brooklyn Museum—is a factor that makes this assignment possible. Other approaches are possible: asking for two models that show separate detail of a single piece in a situation where only one sculpture or other artwork is available could work well. Regardless of those specifics, creating new models using photogrammetry, editing the models in a way that shows agency and decision-making by the student, and then creating a small 3D composition that is uploaded and annotated in SketchFab is the digital means by which students express their understanding of some aspect of the ancient world. I have found that learning takes place during creation. Taking enough pictures of an object so they can make a 3D model affords students the opportunity to look at ancient art closely. That close looking leads to identification of a detail or overall aspect that can be tied to class readings. Bringing it all together into a 3D composition that is accompanied by some writing requires care to do this well. As I have said, I hope to have described this process in a way that allows other teachers to find their own path to integrating any part of this process and these digital methods into their own courses.

The only further note on this assignment that I will add is that it has become my practice to require students to take a "selfie" of themselves and the objects they are modeling. At first I did this as a small indication that students had actually visited the museum and seen the objects themselves. I have found, however, that students enjoy this step. They often pose with the object in a way that is itself an interpretation. The most successful assignments more-or-less directly integrate this selfie into the written component of their work. This is

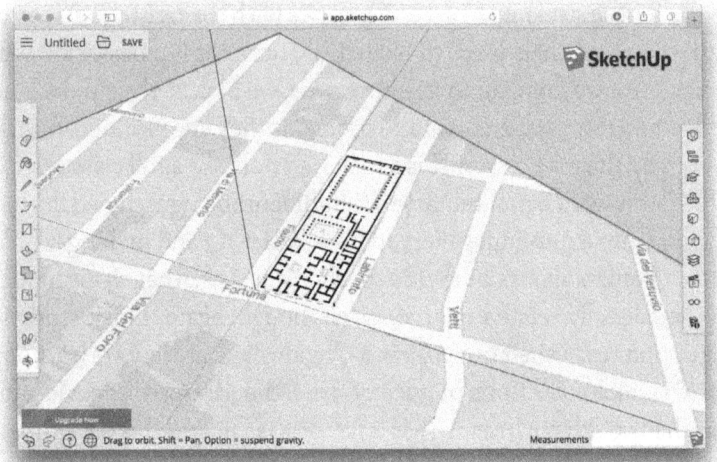

Figure 5: Plan of House of the Faun at Pompeii 'geolocated' in the cloud-based version of SketchUp.

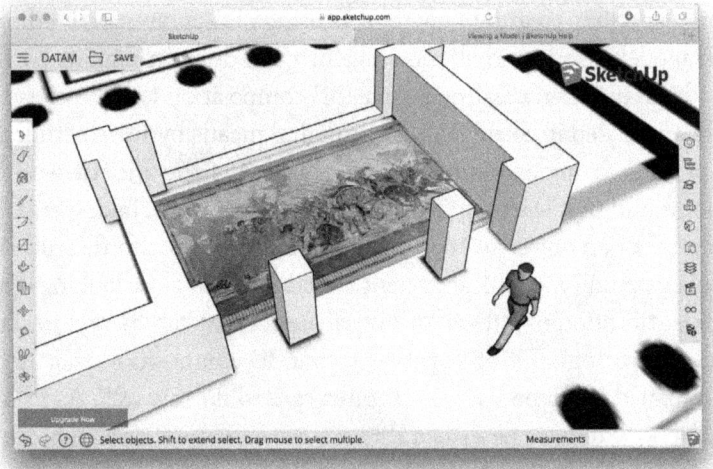

Figure 6: Detail of House of the Faun with basic modeling in SketchUp of location of the "Alexander Mosaic".

fun for me and I am surprised by their creativity.

From 3D Models to Maps

I will move much more quickly through the sequence of assignments and tools that introduce georeferenced data and digital mapping as a tool for learning about the Roman Empire. I introduce the first step in this direction as a natural evolution of the 3D work so far, one that builds on the brief incorporation of evidence from Pompeii that has already come into the class. Using patterns that are familiar from the discussion above, the specific method I first introduce is modeling the walls of houses at Pompeii using the 3D software SketchUp, which is a tool that has been widely adopted in educational settings. Figures 5 and 6 shows the early stages of modeling the House of the Faun at Pompeii (VI.12.2) in the cloud-based edition of SketchUp, which is now the only freely available version of that tool that is kept up to date. While it is the case that I have not taught with this specific version, it probably is what I will use in the future. Two aspects of figure 5 are relevant here. Firstly, the architectural plan, which is available from Wikipedia, is geospatially referenced to a plan of the city. This is accomplished using SketchUp's "Add Location" function. This means that the model could be imported into Google Earth or a Geographic Information System (GIS), though I won't discuss doing so here. As importantly, resizing the plan to fit on the imported street map of Pompeii means that it is to approximate scale. This becomes useful when looking at Figure 6. That shows preliminary modeling of the walls around the space displaying the famous Alexander Mosaic, one of the many artworks from the House of the Faun. I have also imported a figure of a "Roman villager" from SketchUp's so-called "3D Warehouse". That came in at the right scale so that it provides a sense of what it might have been like to walk past that feature. That "sense" is schematic at best so I stress that these techniques can be taught to students in one or two class sessions. The first assignment that results involves preliminary modeling of a Pompeian house, virtually placing mosaics and paintings on the floors and walls, and also adding people. This last aspect asks students to think about which people and how many should be included in a 3D model. How many men, women, and children can a room hold? What roles do those individuals play and

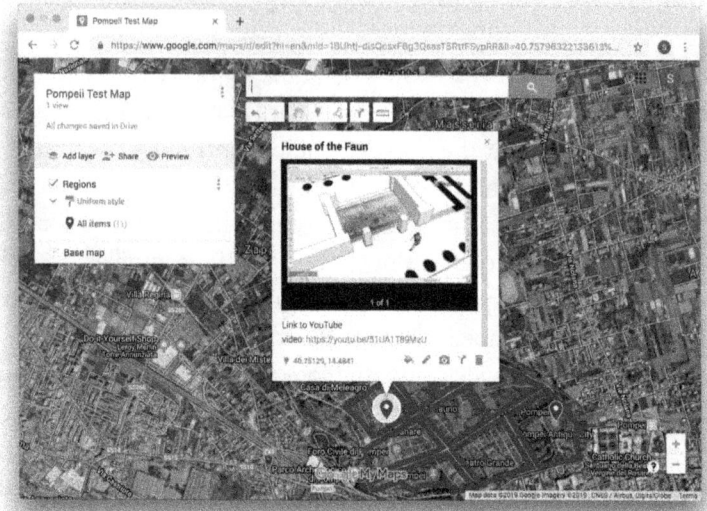

Figure 7: Pompeii in Google Maps with example pop-up at location of the House of the Faun.

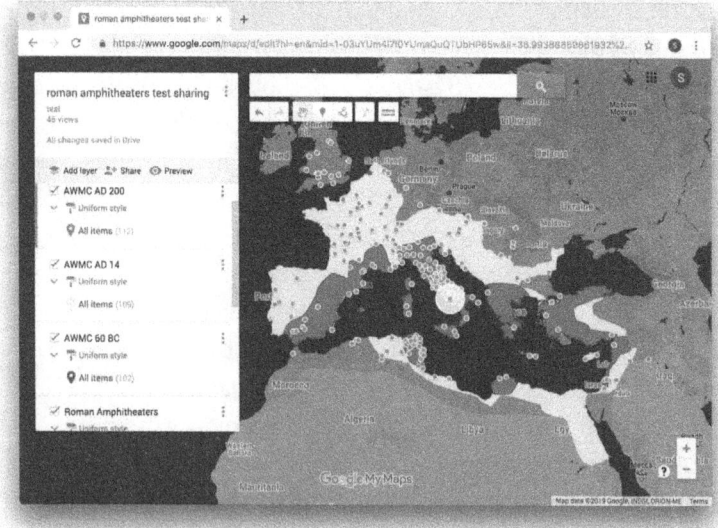

Figure 8: Empire wide map with distribution of Amphitheaters with Pompeii highlighted. Compiled by author from open resources, including shapefiles from the Ancient World Mapping Center.

where in the house are they placed? These are questions that overlap with the digital painting assignment, though here the prompt is the creation of lived-in spaces.[8] In asking students to think about these issues, I have found that the chapters in the volume *World of Pompeii* provide a useful set of readings.[9]

Figures 7 and 8 progressively expand the scope of the mapping techniques I ask students to use and of the maps I ask them to make. They move from the specific house, to the city-wide scale of Pompeii, to the entirety of Roman territory as it expanded from the late first century BCE through the early 3rd Century CE. Figure 7 shows a map created with Google's "My Maps" tool to indicate the location of the House of the Faun in the context of a satellite basemap and an imported KML file that shows modern Pompeian regions. The pop-up shows a screen capture of the SketchUp model that illustrates the specific location of the Alexander Mosaic. When students are constructing their own maps of the city, the website *Pompeii in Pictures* is an exceedingly useful resource, both for locating Pompeian addresses and as a source of plans and imagery.[10] The larger point I mean to illustrate is that content created in one tool can appear in another digital environment. And again, thoughtful choices about what to bring forward into another assignment is an opportunity for students to express their understanding of the Roman world and how digital tools can be used to communicate that understanding. Maps are an effective context in which this communication can happen.[11]

Figure 8 represents another leap. Pompeii is the site of a very early, if not the first, permanent stone-built amphitheater.[12] Its presence provides a segue from modeling and mapping within Pompeii to a wider discussion of amphitheaters and the associated activities of gladiatorial combat, staged hunts, and public executions.[13] Those are all empire-wide phenomena and Figure 8 shows one step in

---

[8] Bozia 2018
[9] Esp. Alison 2007, Bernstein 2007, George 2007; also Alison 2009.
[10] Dunn and Dunn nd
[11] Mostern 2013, Schindler 2016
[12] Welch 1994, Welch 2007
[13] Laurence et al. 2011: Ch. 10

integrating digital mapping into their study. It is again a map created using Google's "My Maps" tool, which means that skills learned mapping Pompeii can be carried forward. Four pre-existing maps have been imported. Three are representations of the extent of Roman territory at 60 BCE, 14 CE, and 200 CE respectively. They are based on ESRI shapefiles created by the Ancient World Mapping center that I have converted into KML files, the format still used by Google Maps and Google Earth Pro. The fourth layer is the result of importing a CSV files that records the longitude and latitude of all known amphitheaters. These resources are all available for download and combining them into a map is well within the capabilities of students. An assignment based on these resources asks students to make active choices about the use of symbols, shapes, and color in creating a map that is itself an effective complement to studying this aspect of the Roman world. As above, when students turn in a map that combines these techniques, I do ask them to describe their work in prose.

Final Projects and Conclusions

I began this paper by suggesting that introducing students to the underlying terminology of 3D models—vertices, edges, and faces—is an appropriate starting point for allowing them to become effective users of digital tools and their place in learning about and representing aspects the Roman empire. I ended with a large-scale map. That is meant to be a linear progression that alludes to a sequence of steps that occurs in class, but which also allows readers to see multiple opportunities to adapt what I have done to their own circumstances. For my students, I use a final project as an opportunity to assess whether or not I have taught them to do just that. Can they define a topic relevant to the Roman world, choose the right digital approaches to represent their work, create a digital resource, and then write about what they have made? Projects have included gathering and creating 3D models of Roman empresses, mapping connections that the port of Ostia has to the rest of the Mediterranean along with modeling the *Piazzale delle Corporazioni*, modeling Roman gardens with Pompeii as a major source, and a map-based presentation of Aeneas' route from Troy to Italy that included ancient representations of elements of that narrative. Each of these projects gave the student attempting it the opportunity to use skills learned in class, some of which I have been able to discuss and illustrate above. And all these projects required engagement with the increasing role of digital data, methods, and tools in Roman studies.[14] Viewing student final projects collectively, I can say that I have been pleased with their authors' ability to create digital content, organize it into a thoughtful digital resource and to complement that effort with a written description of their project and how its digital expression has informed their understanding, and even enjoyment, of the Roman Empire.[15]

List of Downloadable and Cloud-Based Resources

- MeshLab: http://www.meshlab.net
- MeshMixer: http://www.meshmixer.com
- AgiSoft MetaShape: https://www.agisoft.com
- Sets of personal photographs that are useful for introducing photogrammetry: https://drive.google.com/drive/folders/0By0Ma66BERCtanFFZkl3Rmg1OXM
- SketchUp for Web: https://www.sketchup.com/products/sketchup-for-web
- Google "My Maps": https://www.google.com/maps/d/ (Note: the exact path to access this tool has often changed and can depend on the specifics of your institutionally provided Google account, if that is what you are using.)

---

[14] Bagnall and Heath 2018
[15] Perry 2019

Works Cited

Alison, P.
   2007 Domestic Spaces and Activities. In *The World of Pompeii*, edited by J. Dobbins and P. Foss, 269-278. Routledge, New York.
   2009 Understanding Pompeian Household Practices through their Material Culture., *FACTA: A Journal of Roman Material Culture Studies* 3: 11-32.

Bagnall, R. and Heath, S.
   2018 Roman studies and digital resources. *Journal of Roman Studies 108*: 171–189. https://doi.org/10.1017/S0075435818000874

Bernstein, F.
   2009 Pompeian Women. In *The World of Pompeii*, edited by J. Dobbins and P. Foss, pp. 526-537. New York, Routledge.

Bond, S.
   2017a Whitewashing Ancient Statues: Whiteness, Racism and Color in the Ancient World. *Forbes*. Available online https://www.forbes.com/sites/drsarahbond/2017/04/27/whitewashing-ancient-statues-whiteness-racism-and-color-in-the-ancient-world/.
   2017b Five New 3D Models of Ancient Artifacts that are Changing How We Interact with Museums. *Forbes*. Available online. https://www.forbes.com/sites/drsarahbond/2017/05/19/five-new-3D-models-of-ancient-artifacts-that-are-changing-how-we-interact-with-museums/#404f85b21e30.

Bozia, E.
   2018 Reviving Classical Drama: Virtual Reality and Experiential Learning in a Traditional Classroom. *Digital Humanities Quarterly* 12(3). Available online: http://digitalhumanities.org:8081/dhq/vol/12/3/000385/000385.html.

Brooks K. R.
   2017 Depth Perception and the History of Three-Dimensional Art: Who Produced the First Stereoscopic Images? *i-Perception* 8.1: 2041669516680114. https://doi.org/10.1177/2041669516680114

Dunn, J. and B. Dunn
nd. *Pompeii in Pictures*. Available online: https://pompeiiinpictures.com.

Garstki, K.
2017 Virtual Representation: The Production of 3D Digital Artifacts. *Journal of Archaeological Method and Theory* 24(3): 726-750.

George, M.
2007 The Lives of Slaves. In *The World of Pompeii*, edited by J. Dobbins and P. Foss, pp. 538-549. New York, Routledge.

Kleiner, D.
2005 Semblance and Storytelling in Augustan Rome. In *The Cambridge Companion to the Age of Augustus*, edited by K. Galinsky, 197-233. New York, Cambridge,. https://doi.org/10.1017/CCOL0521807964.010.

Laurence, R., S. Cleary, and G. Sears
2011 *The City in the Roman West: c. 250 BC - c. AD 250*, Cambridge, Cambridge University Press.

Mostern, R.
2013 Traveling the Silk Road on a Virtual Globe: Pedagogy, Technology and Evaluation for Spatial History. *Digital Humanities Quarterly* 7(2). Available online: http://www.digitalhumanities.org/dhq/vol/7/2/000116/000116.html.

Olson, B., R. Placchetti, J. Quartermaine, A. Killebrew
2013 The Tel Akko Total Archaeology Project (Akko,Israel): Assessing the suitability of multi-scale3D field recording in archaeology. *Journal of Field Archaeology* 38(3): 244-262.

Perry, S.
2019 The Enchantment of the Archaeological Record. *European Journal of Archaeology* 22(3): 354-371

Rabinowitz, A.
2015 The Work of Archaeology in the Age of Digital Surrogacy. In *Visions of Substance: 3D Imaging in Mediterranean Archaeology*, edited by B. Olson and W. Caraher, pp. 27-42. Grand Forks, ND, The Digital Press at the University of North Dakota..

Schindler, R.
2016 Teaching Spatial Literacy in the Classical Studies Curriculum. *Digital Humanities Quarterly* 10(2). Available online: http://www.digitalhumanities.org/dhq/vol/10/2/000252/000252.html.

Watrall, E.
2019 A New Approach to Digital Heritage and Archaeology. *Cambridge Core blog*. Available online: https://www.cambridge.org/core/blog/2019/05/21/a-new-approach-to-digital-heritage-and-archaeology/.

Welch, K.
1994 The Roman Arena in Late-Republican Italy: A New Interpretation. *Journal of Roman Archaeology* 7: 59-80.
2007 *The Roman Amphitheatre: From Its Origins to the Colosseum.* Cambridge University Press, Cambridge.

Woolf, G.
2018 Strangers in the City. In *Instrumenta 64 Xenofobia y Racismo En El Mundo Antiguo*, edited by F. Marco Simon, F. Pina Polo, and J. Remesal Rodríguez, pp. 127-136. Barcelona Universitat de Barcelona Edicions, Barcelona.

Digital Janiform: The Digital Object from Research to Teaching

Eric Poehler

The primary argument of this paper is that the digital object—in this case defined as a product of the dual engagement of objects of inquiry and digital technologies—can face toward research or teaching equally and without contradiction. I argue further that the digital objects that describe the ancient Mediterranean world are produced commonly in the context of research, but not commonly enough reoriented toward teaching. In what follows, therefore, I discuss my own attempts to bring digital objects that I have produced or have received access to in the course of research into my classes. These include attempts to incorporate video tours of ancient cities to interpret trench notebooks and data records from excavations (Part I), to digitize and map artifacts of Pompeii's early modern excavators, and to measure the impact of cart flows inside a 3D scan (Part II). A final section of the paper looks at the relationship of students and faculty to digital objects in the classroom, examining both the readiness of students to manipulate these materials and the faculty member's capacity to deploy them. While Part III might naturally appear first in the discussion and frame it, I have left it in chronological order so that the lessons therein can reflect back on the examples described in preceding sections rather than appear, prescriptively, to explain them.

Part I. Technology: from Fieldwork to Classroom.

Video Tours of Ancient Cities

In the summer of 2015, as part of a search for evidence of the circulation of traffic in the Roman cities, I visited a number of sites in the southwest of Turkey and the southeastern coast of France. My goal was to identify, record, and analyze the wearing patterns on street features that were indicative of the direction of traffic on a given street.[1] In support of this, I wore a Google Glass headset and recorded an approximately hour long tour across each ancient city, locating and recording the context of the evidence for traffic during the tour. Initially, my idea was that the video capability of Google Glass would allow me to passively document much more visual information about the evidence for traffic including its broader spatial context. It became clear almost immediately, however, that it was possible to record much of the experience of visiting the ancient city, including the character and shape of the excavated city.

What took longer to understand was that these videos were of comparative value, and, if viewed in a particular order, they had the potential to show students the varying landscapes of Roman urbanism. Even for the passive viewer, the differences in geography and geology (where in the empire the city is and what building stones are available), chronology and scale (how old and how big the city is), and levels of preservation and excavation (how a city was destroyed and how it was recovered) serve to make each city a unique visual experience. With sufficient time, simple exposure can produce a kind of iconography of a site, a comfort, a *habitus*, that manifests in a feeling that you know generally what to expect around each new corner. For more than 250 years, Pompeii has served as the quintessential archetype of the Roman city. It has been presented as both the ideal (the best preserved and the most excavated) and the norm (located in the heart of the Roman world and its history). Although neither is fundamentally true, this popular expectation makes the visual experience of Pompeii a kind of "control" experience for comparison with other Roman cities.

With this in mind, I created another Google Glass video of Pompeii and placed it at the start of an assignment designed to create for students something of the comparative urban experience I had by actually visiting the sites.[2] To begin, students first watched the video

[1] Results of this survey can be found in Poehler 2017: 216-24.
[2] See Addendum 1: Assignment – Late Roman Cities – Ancient Material Worlds. (https://archive.org/details/Poehler_DATAM_Addendum_1/) The PDF version of this addendum contains the text of the assignment as well as links to the videos

tour of Pompeii, taking notes and building up their "control iconography" that they would use to find similarity and difference in the other cities they would view, one in Turkey and one in France. Students were also provided with an online GIS map with satellite imagery of each city and given basic instruction on using a simplified version of this technology. To complete the assignment, students were given three requirements. The first was to draw a line on the map indicating the path the video took across the city. This ensured they watched the entire video. To make their engagement more active, students were asked to identify five important landmarks, buildings, or features along this path and place a digital "pin" at each location. Within this marker they were required to write a 100 to 200-word description of the place, including citations to their sources of information and to other imagery that could be linked from the web. Finally, at the end of their three tours, students were asked an open-ended question to compare and contrast these urban landscapes in a not-less than 1000-word essay.

Reconstructing Excavations from Digital Fieldwork Records

In the popular imagination, excavation is synonymous with archaeology. For archaeologists, of course, it is only one—albeit well-represented—of the many methods by which we explore the past. Teaching excavation is notoriously difficult to accomplish in a classroom rather than in the field. One common classroom approach is to deconstruct the process of excavation and atomize the method into its component principles, giving lectures or assignments that

---

("S" - stabilized version; "US" - original, unstabilized version) and the map of each site. The YouTube playlist for all stabilized video is here: https://www.youtube.com/playlist?list=PLWZ6-0WL9ynUrd0A97P6_bTqev2PwkAym.
The YouTube playlist for all unstabilized video is here: https://www.youtube.com/playlist?list=PLWZ6-0WL9ynUBaWOS4hY-9rmem11EwW3U. Links to each map are as follows: Ambrussum (https://arcg.is/1nLO4P); Aspendos (https://arcg.is/1bDGre); Ephesus (https://arcg.is/1PbH4q); Glanum (https://arcg.is/19fG5j); Hieropolis (https://arcg.is/1HmvHO); Laodiceia (https://arcg.is/0qfPnX); Perge (https://arcg.is/iOzfW); Sagalossos (https://arcg.is/04u9ba); Side (https://arcg.is/1e10my); Vaison-La-Romaine (https://arcg.is/1PiaOT).

define (for example) stratigraphy, seriation, and *terminus post quem / terminus ante quem*. But in doing so, at least in my own experience, we disconnect the principle from the practice such that, in the end, students are well prepared for a test, but not for a trench. They can define stratigraphy, but they can't recognize it. Moreover, the atomization of excavation into its component principles followed by its recombination for students in a single example also regularly projects the notion that real-world excavation is equally regular and that archaeological practice is standardized and uniform.

This is not the experience one gets from working in the field and certainly not from working on multiple projects. My own experience of contributing at different levels on three different projects at Pompeii, a decade in the field and in the archives at Isthmia, Greece, and a brief stint at Morganitina, Sicily has taught me slowly (and largely passively) that archaeological practice varies greatly over time (e.g., between 1970 and 2009), across subdisciplines (e.g., between Greek and Roman archaeology), and even among trenches within the same project (e.g., the particularities of supervisors in recording practices). Although my experience matured in a long-term interaction with the data and the information structures that each project deployed—from trench notebooks at Isthmia,[3] to maps of walls labeled with the Pompeii Forum Project's particular nomenclature system,[4] to single context recording forms and digital notebooks used by the Pompeii Archaeological Research Project: Porta Stabia[5]—this greater conception of my understanding of excavation, however, only materialized when trying to figure out how to teach it to others. Still, coming to this understanding also convinced me that if knowing the multiple ways we conduct and record excavation could be won by accident of long term exposure, it should also be possible to compress that experience into a classroom setting using the same digital objects in order to produce a similar understanding.

---

[3] These notebooks have been digitized by Jon Frey and the ARCS project: http://dev2.matrix.msu.edu/arcs/projects/single_project/isthmia

[4] See below map: http://pompeii.virginia.edu/forummap.html

[5] Ellis 2016.

Therefore, after a day of lectures about the component principles of excavation mentioned above, students in my Roman Archaeology course embarked on an investigation of the digitized trench notebooks from Isthmia, Greece. Each student read one of five notebooks from 1970, 1971, or 1972, and then answered a series of questions in a Google form.[6] These questions, at first, were seemingly simple. For example, "how many trenches were recorded in the notebook?" and "what are the dimensions of the trench(es)?". These questions were designed to reveal to the student the size of the area and the pace of work the archaeologists named in the notebooks were supervising. Other questions appeared even more innocuous at first sight, including "what is a basket?" and "what is a box?", but were in fact especially challenging and important questions. Asking students to identify the primary units of recording in the trench notebooks not only challenged students to understand the terms excavators used within their project, but also to evaluate the degree to which these practices reflect an adherence to the stratigraphic principles they learned in theory. Identifying, naming, and numbering is the most basic act of the field archaeologist, but it also necessarily reveals her most basic values; illegible coins are traced as circles in these notebooks, while the relationship of one deposit of soil to another is left implicit by the order of recording. The final questions in the Google form were more impressionistic, asking students to share what was most interesting or confusing about the notebooks. These questions gave student a place to express their curiosity and vent their frustrations, and also served as a conversation starter for the next class meeting.

Part II of this exercise turned from digitized notebooks, to a born-digital excavation record from Pompeii, generated four decades later. This digital record included both the trench supervisor's notebook and well as the 144 database forms that documented each

---

[6] See Addendum 2: Assignment – Excavation – Ancient Material Worlds (https://archive.org/details/Poehler_DATAM_Addendum_2/). The PDF version of this addendum contains the text of the assignment as well as links to trench notebooks of Isthmia used. The material for Pompeii, however, is not able to be shared. At the end of the assignment, however, copies of the Google forms for both Isthmia and Pompeii are included.

stratigraphic unit as it was encountered. Armed with these digital data, students were asked to document what happened on each day during one week of the excavation, reporting how many and what kinds of stratigraphic units were discovered, the finds and features that were highlighted, and (again), what was interesting and confusing. The simplicity of this process of reading and reporting was intended to generate familiarity with the process and the data at Pompeii and, by comparison with the notebooks from Isthmia, to make room for observations about the variability of practice within excavation as a method. The first such observation was that scale and speed of the excavation had shifted dramatically; trenches were much smaller, dug more slowly, and the efforts of recording were expended on soil stratigraphy and features rather than artifacts. In fact, at Pompeii, the segregation of excavators and finds specialists was almost complete. Conversely, another difference for students to recognize was how the introduction of digital technologies in the trench distributed the recording process to all members of the team, rather than being the responsibility of the supervisor alone.

These are only a some of the questions that might be asked from these digital objects, only a few of the ways in which they might be arranged in an assignment, and importantly, these are only some of the digital objects that are available, ready to be incorporated into different exercises. Beyond this flexibility, the value of the digital object is to authenticate the learning experience by placing real-world products, the outputs of academic fieldwork, into the hands of students without the intermediating structures of an overly reductive lesson. But because these data were not carefully packaged, the assignments built from them must be general and simple. In this way students learn in the classroom the way they learn in the field: by contact with its materials, by repetition of its procedures, and in recognition that there is no right answer. The hope is that students are building up an iconography of an excavation, learning its principles (e.g., stratigraphy), the forms of its data and information structures (e.g., notebooks and context sheets), as well as its general (e.g., "basket", "stratigraphic unit") and its specific nomenclatures (e.g., "marl", "stereo", "lapilli", "quarry pit").

Part II. Technology: from Exercise to Collaboration

If the reality of the digital object in assignments about excavations can begin to approach the experience in the field, one crucial element that creates that authenticity remains missing. However real the digital object is, those trenches are closed, those projects have ended, and the assignment their data are used within is still just an exercise. What makes fieldwork engaging is the tension among the physical effort of the work, the anticipation of discovery, the fear of making a mistake, and the chance to make a genuine contribution to the history of the ancient Mediterranean world. By giving a student a stake in the outcome of a learning opportunity, we create the opportunity to push beyond authenticity to a sense of ownership, which can have implications beyond the exercise and the class for both the student and the data. I have attempted to create this opportunity in two of my assignments by creating, analyzing, and interpreting digital objects from Pompeii.

Creating and Mapping the Earliest Artifacts at Pompeii

The data that are available from the Pompeii Bibliography and Mapping Project (PBMP) have been a boon to both my research and my teaching.[7] Because I direct that project, it is easier for me not only to access and implement those data compared to other users, but also to imagine novel uses and opportunities. One major gap in the PBMP is the absence of the artifactual record, which leaves the user of the online map beholden to architectural features and categories to understand the shape and use of Roman urban space. Fortunately, Mario Pagano and Raffaele Prisciandaro published a listing of all 3,253 objects recorded in the daybooks between 1748 and 1859 in an appendix to their landmark work on the early excavations at Pompeii.[8] The opportunity to add some of the missing artifacts back into the

---
[7] http://digitalhumanities.umass.edu/pbmp/
[8] Mario Pagano and Raffaele 2006

urban landscape now became a reality and along with it, the opportunity to help students understand the impact of changing archaeology practice on our perception of the archaeological record.

Over the course of two classes, one in 2014 and another in 2016, we first digitized and later mapped the data in Pagano and Prisciandaro's appendix, which contained information about the date an object was found, its type, material, as well as information about where an object was published (in case of inscriptions) and/or where it was currently held in a museum collection. By digitizing these records, the 2014 class was able to do simple, yet unprecedented calculations to approach basic questions, such as "What kinds of artifacts did the early excavators privilege most?", "Did they collect more objects over time?", and "Did they collect a greater variety of objects over time?". With these new digital data that they created, students were also able to make informed speculations about those things we could not calculate, such as "Why did the early excavators not record many pottery finds?" and "Where are the many coins that must have been encountered?". These questions and the answers suggested by the data served as an ideal means to introduce the concept of site formation processes and to address the so-called "Pompeii Premise", which is the notion that Pompeii was a site frozen in time and therefore a perfect record of the ancient Romans' daily life. By creating their own archaeological records in a digital format, students were able to tackle an important subject, which had not yet been approached quantitatively.

In 2016, I taught the Pompeii class again and we returned to the Pagano and Prisciandaro data because, importantly, their appendix contained the approximate location of where each object was found. Ironically, because the spatial information was an address to an entire building rather than a particular room or a specific find spot, it made mapping these data easier for students who were only beginning their familiarity with Pompeii's entire urban landscape. Fortunately, my institution has a deep investment in ESRI's ArcGIS software and its accompanying cloud services, ArcGIS Online, which allowed me to give each student their own account to map artifacts using only a web browser. Still, students of course required training in the platform and software, so we dedicated a day to learning the basics of

ArcGIS Online and the process to map artifacts, which I recorded as a screen-capture tutorial and posted for the students to revisit at any time. After the students had placed a digital pin for each object inside its appropriate property, I joined the table of data digitized from the 2014 class and students were asked to explore some of the questions that were impossible to ask in the previous class. Because of the digital objects they created and the mapping environment they could work within, these students were able not only to examine the spatial distribution of statues, graffiti, and carved gemstones, but also to consider those distributions against the types of buildings in which they were found—houses, shops, workshops, and public buildings.

Indeed, before this mapping work, it was impossible for any scholar to approach these questions. These students were the first people to look at these data in these ways, to slice through them both chronologically and spatially, and were the first to try to understand their meaning. Their efforts were real and genuine forms of primary research. Within the classroom, these new digital objects afforded a unique opportunity to consider the degree to which our artifactual record was skewed by the priorities of the early excavators and by the flight of desperate Pompeians 1,900 years ago. All together, students produced the information they needed to explore more of Pompeii's urban topography and the landscape of human choices that now influence how we can interpret that urban form. What is more, these students made a real contribution to the subject they supposed they were only meant to be learning about.[9]

Analysis of a 3D Scan of Axle Wear at Pompeii

During fieldwork at Pompeii in the summer of 2016, I took a relatively new technology called the "Structure Sensor" into the ancient city's streets in order to capture in 3D some of the textures of those environments: the paving stones and the ruts, curbstones and sidewalks, the drains out from houses, the ramps leading into inns, and

---

[9] Data from the class is hosted as a layer—"Artifacts of the Bourbon Excavations"—on the Pompeii Bibliography and Mapping Project's online map: https://www.arcgis.com/home/item.html?id=a932a86e11ba4ba28eabfa5976cec33b

the hitching holes cut in for tethering animals. The results of these informal and exploratory attempts were rough, but nonetheless they recorded more information and in more detail than I had ever done with pad and pencil in nearly two decades of surveying those same streets. In one of the scans, a long scar along a raised curb showed the place where carts had come into contact with the soft travertine curbstones. It occurred to me that, as can be done by measuring the position of pairs of ruts, it might be possible to discern something of the size and construction of the vehicles that made these wearing patterns. It also occurred to be that bringing these digital objects into my Pompeii class—untouched on unexamined—could bring alive the experience of being in Pompeii's streets and create an opportunity to learn not only what is already known about ancient traffic, but also to make a new contribution to that discussion as well.

Once again, the work asked of the student was intended to be straightforward and to mimic working in the field: make a series of measurements (within the 3D model), record those measurements (in a custom built spreadsheet), and finally comment on the initial results.[10] To do this, it was necessary to make some affordances for the technologies involved. For example, although the data came from the structure sensor as complete model files with textures, few students could be expected have software capable of manipulating and measuring within those models nor a background in using such software. Fortunately, a 3D model can be embedded within a PDF and the basic Adobe Reader program has a suite of tools for interacting with those models. In addition, many students were unfamiliar with using 3D models in general and it was necessary to give a tutorial in class to introduce the concept of the assignment, the 3D models, and the

---

[10] See Addendum 3. Assignment – Curb Wear and Cart Design – Ancient Material Worlds (https://archive.org/details/Poehler_DATAM_Addendum_3/). The PDF version of this addendum contains the text of the assignment as well as links to SketchFab where the 3D models used in the assignment are hosted and to each model embedded within a pdf. At the end of the assignment a copy of the spreadsheet used to tabulate the results of measurements within the models is included. Finally, a brief tutorial video for the assignment is here: https://youtu.be/yypqkkC1u2c.

software we would be using. As with the assignment on Pompeian artifacts, I recorded this tutorial with a screen capture software and linked the video to the assignment materials so students could return to the discussion when necessary.

In the end this was a more exciting, but also a more frustrating and less successful assignment for the students. Part of this was technical. Although embedding the 3D modeling into a PDF will make the model easily accessible and shareable, it was difficult to move the model while taking a measurement, which resulted in parallax errors. That is, they could see a measurement to make from one perspective, but when they rotated the model again, one end of that measurement was as out of plane with the other, making the dimension measured incorrect. Naturally, having to redo the work was frustrating for students and made the results less consistent and credible.

The assignment was a pedagogical experiment, and like many experiments, this one failed to reach its stated goal but achieved another nonetheless. By admitting the failure of this project, an equally interesting discussion arose in class about how such setbacks require one to reconsider the original project's design and make improvements. In these ways, we decided, failures lead to advances in archaeological method even as they lead to delays in writing archaeological history. There are also lessons for designing assignments that lean on digital objects and student collaborators. The first is to not over conflate one's own comfort and abilities with the technology or the subject matter with that of the supposed average student. As we will see, the average is likely less technically inclined than some might expect. A second lesson is about setting the right tone for the assignment and expected outcomes. When confronted by an un(der)-defined outcome, students naturally become anxious over the prospect of a poor grade, a reality that only amplifies their frustration when aspects of an assignment impede their efforts to do good work. In this one assignment, at least, I found that by regularly reminding students that the assignment was an experiment, that experiments can fail, and that failure is often the answer to a different question, the anxiety of a potential poor grade could be diminished and many students could explore the assignment with a freer sense of mission, one less tethered to a grade. Of course, a few less committed students took this rhetoric as an opportunity to not try very hard, and, in that, succeeded.

Part III. Technology: from Students to Faculty

At the time of writing, I have served as the director of the combined programs of Blended Learning and Digital Humanities for the Five Colleges for a year and a half. My initial duty was to collapse these two, large-scale, formerly Mellon Foundation-funded initiatives into one joint program and it was this charge that brought the idea of the janiform digital object to mind for me for the first time. As we closed the independent Blended Learning program and wrote our final report for the Mellon Foundation, we decided to survey the thirty-one faculty members who had led projects under the four-year grant about their experiences teaching with technology. At the same time, I was also about to begin teaching a first-year seminar at my home institution, the University of Massachusetts Amherst, called "Digital Tools and the Academy". This course is a primer of basic technologies currently used across the campus and that will likely be used over a four-year academic career. To get a sense of what students entering university for the first time knew about academic technologies, I asked them to fill out a survey of their experience and comfort with a range of technologies. Together, these two surveys—what students know about technology when entering university and what faculty exiting a pedagogical project think about technology in the classroom—bookend the experiences of Blended Learning and offer some anecdotal information about future best practices.

Technologies of First Year Students

On the first day of class, after an introduction to the course, I presented students with an online survey called "What we Know Now". In this survey, students were asked first about their basic levels of comfort with technology, to which 81% reported being "comfortable" or "very comfortable". No student reported feeling "very uncomfortable" with technology. Next, the survey asked about each of the basic suite of digital tools in Microsoft Office and Google Apps. The scale was intentionally casually worded and ranged from "I develop for it", to "I use it almost daily" to "I use it", to "I used it once", to "I know it's a thing" to "Huh?". The results were perhaps unsurprising: nearly all students knew about Microsoft Word and PowerPoint; some were familiar with Excel, but Microsoft Access was almost entirely unknown to them. Google products were even more familiar. In this group, every student used Google Docs and almost every student used it daily. Both Google Sheets and Slides are more commonly used than their Microsoft counterparts and even Google Forms were known to a few students. Conversely, when asked later, none of the students recognized that the online survey they had taken was built in Google Forms. Finally, only Google Drawings were not well known; during our in-class discussion students reported using Microsoft Paint instead.

The survey also asked students about two other types of technologies, which I termed "Academic Technologies" and "Digital Platforms". Once again, even with email and cloud storage services equally offered via UMass, students were most aware of Google products and used GMail and Drive over Microsoft Exchange and Box. Dropbox was the next most popular cloud storage service. Perhaps the most surprising and important result of the survey was the lack of knowledge and confidence that the students had with using library technologies. Although the lack of familiarity with citation management programs is to be expected, no more than 15% of this group of first-year students reported using the library search tools or library databases more than once. Interlibrary loan was completely unknown. There was a similar lack of awareness with the digital platforms used at UMass, which they would encounter in the course of their academic careers. Hosting solutions were unknown and even Wordpress, the internet's dominant content management system, was only used once by one student. Instead, a plurality of students were regular users of audio and visual processing software.

These limited results paint a picture of the levels of knowledge of and comfort with technology that students might hold as they enter the university. Moreover, these results suggest that there are some pre-established areas of familiarity that instructors might lean upon if they want to bring digital objects into the classroom. At the same time, it is clear that there are many areas of technical knowledge that students do not possess and sufficient time must be devoted to the development of that knowledge should an digitally-infused assignment require it.

Technologies of Blended Learning faculty

In teaching, there are many neologisms that a faculty interested in bringing digital objects into the classroom will be confronted by: Blended Learning, Distance Education, Hybrid Learning, and MOOC, to name only a few. Many of these terms are used interchangeably, while at the same time they have correlations with the different challenges that different higher education institutions face. For the Five Colleges consortium, consisting of Amherst College, Hampshire College, Mount Holyoke College, Smith College, and the University of Massachusetts Amherst, the challenges are not overcoming physical distance or asynchronous course schedules. Instead, for these institutions who pride themselves on high-quality, face-to-face teaching in small group settings as well as opportunities to participate in world-class research, the question in introducing digital objects into the classroom is how does one improve the student's experience without jeopardizing the core of current practice and success?

To begin to answer that question, the Five Colleges Inc. sought and received a major grant from the Mellon Foundation in 2014 to support innovation in teaching with technology in the humanities and humanistic social sciences. Our program sought to find multiple points of entry into these curricula, but did not desire to see any class transformed to become entirely online. Thus, we wanted to seamlessly blend new technologies and traditional teaching, naturally becoming the Five College Blended Learning Program or 5CollBL. Between 2014 and 2018, we supported 31 faculty in 24 courses. In

2017, Blended Learning was combined with the earlier and very successful Digital Humanities program to form a single joint program. Over the summer of 2018, when our funding for Blended Learning was exhausted and our final report to Mellon was coming due, we produced a very short survey for our faculty to get a sense of their experience of Blended Learning. Unlike the survey of first-year students, this survey was voluntary. Nonetheless, we had a 97% response rate.

In the survey, we asked faculty to rank the following questions from "very negatively" to "very positively":

- How did your experience with the Blended Learning project impact your teaching?
- How, on average, did students react to the Blended Learning technologies used?

To the first question, faculty responded with an overwhelmingly positive opinion of the impact of technology in the classroom on their teaching. Faculty perceptions of their students' feelings about the technologies used were similarly positive. Of course, the respondents were a self-selected group of grant-funded faculty, and so too, to some degree, were their students. Therefore, it is unsurprising to find some high opinions of Blended Learning practice in general. When we asked faculty to rank, between Completely and Not at All, "How much did your Blended Learning experience impact your other courses and teaching?" the results were mixed. Most faculty reported a neutral or limited effect, though only a handful said it had no impact on their other courses.

The final two questions were open ended responses and the comments faculty shared help to underscore the position of the digital object in the minds of many faculty teaching in humanistic disciplines today. In the first instance, these faculty expressed that, at the outset, the amount of work to integrate digital objects and technologies into the classroom is significant. Several believed it would not have been possible, or at least not undertaken, without the financial support from the Mellon Foundation's grant to the Five Colleges.

In my own experience in the program as an awardee rather than as its director, what most facilitated the work was the existence of digital materials I had already created in the context of research. Had I been required to generate both the digital content and the instructional context for that content (as some of my colleagues did), I am not sure even the financial incentives from the grant would have been sufficient. On the other hand, if the barrier to entry is high in Blended Learning, the ease of reuse is equally high. Many of our faculty reported that the work done under the grant made offering and reworking similar assignments in subsequent courses simple. Above all, faculty thought that Blended Learning was a useful addition to their pedagogical repertoire and believed that it was a useful part of the future of university education, though not *the* future. One instructor summed it up this way: "It is a nice enhancement, but not a substitute for face-to-face teaching." Teaching in the liberal arts can be meaningfully enhanced, but small-scale, in-person dialog continues to be perceived as the most effective and most rewarding mode of instruction.

Conclusions

The preceding sections have described some of the different ways I have attempted to use digital objects in teaching the ancient Mediterranean world and in what follows I will sum up some of the lessons learned in those attempts. In the first instance, it seems clear that video from archaeological sites and 3D models of objects and excavations can genuinely enliven a student's perception of the ancient world. By simply being more visually available, students can begin to build up an iconography and familiarity, a sense of place and landscape that professionals gain from spending days and even years on site. Similarly, putting the actual records of excavation into the hands of students—even with all the rough-edged realities those records entail—can bring a more genuine sense of what digging in those trenches was like.

Indeed, in teaching a method like excavation, those rough edges within the digital objects can actually be a benefit. Rather than imperfections that need to be smoothed out in a carefully curated assignment to explain a concept or principle (e.g., stratigraphy), these vagaries of the data can be repackaged to demonstrate that archaeology is not a single, standardized process, but a multiplicity of practices across decades, projects, even among trenches. Such "imperfections" provide an opportunity to discuss issues of research focus, of methodological choice, and how data are created and their consequent incommensurability. To do this requires the instructor to create an environment of openness toward the process: to reward a flexibility of mind and to tamp down the common transactional expectation of learning fact and passing an assessment. In my experience, few means of encouragement serve better to create that environment of openness in the classroom than inviting students into your digitally-based research projects and into your trust. Students feel a sense of investment in a project when they know a project has a real-world research outcome, when they know the instructor has a stake in the result, and when they know they will be the first people to ever see the past revealed in a new map, a new graph, and a new table of data.

Finally, although such assignments do not need to be narrowly built to lead students through the material to the correct outcome, they do require a clear-eyed assessment of the technological barriers to success. Our survey of incoming students indicates they come to higher education with a core, but limited core, of digital skills. They are not the digital natives some imagine them to be. Therefore, additional training in the technologies used in assignments must be introduced deliberately, but also with a sense of the simplicity of the process that they are asked to accomplish. Likely, the software being used can do much more than the assignment requires and students will benefit from knowing they only need to learn a small part of it. Nonetheless, the instructor will need to make new training materials, dedicate class time, and/or arrange for institutional support personnel to help.

All of this work takes time, lots of time for the initial course preparation. Nonetheless, our survey of faculty members who have built such digitally infused courses from scratch report it to be a rewarding and transformative process. Within the classroom there is often a renewed sense of purpose. Beyond the classroom, faculty see a new and clear connection between previously competing components of their labor: research and teaching. The benefits accrue further in later semesters as reusing materials, in the short term at least, requires little new effort. Within a decade, however, we should imagine updates to online platforms, file formats, access portals, and institutional support regimes will continually diminish the return on the initial investment in the course until it becomes unviable. While some might see this as a failing of the technology, I would argue that ten years is the life cycle of any assignment, as learning styles, institutional expectations, and (not least) the underlying content equally change to make it obsolete. Fortunately, if we are continuing to research in areas we teach, the janiform nature of the digital object should continue to permit us to meaningfully reorient our research data toward teaching materials and continue the virtuous circle.

Works Cited

Ellis, Steven
   2016 Are We Ready for New (Digital) Ways to Record Archaeological Fieldwork? A Case Study from Pompeii. In *Mobilizing the Past*, edited by Erin Averett, Jody Gordon, Derek Counts, pp. 51-75. Grand Forks, The Digital Press at the University of North Dakota.

Pagano, Mario and Raffaele Prisciandaro
   2006 *Studio sulle provenienze degli oggetti rinvenuti negli scavi borbonici del Regno di Napoli: una lettura integrata, coordinata e commentata della documentazione.* Nicola Longobardi, Castellammare di Stabia.

Poehler, Eric
   2017. *The Traffic Systems of Pompeii.* New York, Oxford University Press.

Contributors

Marie-Claire Beaulieu is a classicist interested in the relationship the ancients had with their environment and the mental constructs they associated with natural phenomena and animals. Her recent book, *The Sea in the Greek Imagination* (University of Pennsylvania Press, 2016), explores the Greek mythological representation of the sea as a space of transition between the living, the dead, and the gods. As the co-director of the Perseids Project since 2013, Marie-Claire has engaged in many interdisciplinary teaching initiatives, in particular with the Departments of Religion and Computer Science, and her classes make frequent use of technology to involve students in the process of research.

Sandra Blakely holds her doctorate in Anthropology and Classics, and is Associate Professor of Classics at Emory University. Her research interests include the archaeology of ritual, digital approaches to the ancient world, and comparative ethnography; she is completing a monograph on the maritime promises of the mystery cult of the Great Gods of Samothrace.

Anthony Bucci is a computer scientist, entrepreneur, and lecturer interested in coevolutionary algorithms and more recently natural language processing. He has served as a scientific advisor or chief scientist and cofounder at multiple startups, coauthored eight articles about computer science education, and co-taught a computational methods class for the last three years with the Classics Department at Tufts University. He is also an active participant in Tuft's Birds in Ancient Mythology project (https://sites.tufts.edu/ancientbirds/), where he explores extracting structured knowledge from the text of D›Arcy Thompson›s *A Glossary of Greek Birds*.

Patrick J. Burns is a Postdoctoral Fellow at the Quantitative Criticism Lab (UT-Austin) working on computational literary criticism. He is also a Research Associate at NYU's Institute for the Study of the Ancient World and contributes to the Classical Language Toolkit, an open-source project offering text-analysis support for historical languages.

William Caraher is an associate professor in the Department of History at the University of North Dakota. He specializes in field archaeology, Early Christian and Byzantine architecture, material culture and settlement in the Bakken oil patch of western North Dakota, and the history of Late Antique Cyprus and Greece. He's the co-author of *Pyla*-Koutsopetria *I: Archaeological Survey of an Ancient Coastal Town* (with David Pettegrew and R. Scott Moore, American Schools of Oriental Research, 2014) and *The Bakken: An Archaeology of an Industrial Landscape* (with Bret Weber, North Dakota State University Press, 2017. He currently edits the literary journal *North Dakota Quarterly*.

Helen Cullyer is a classicist, who has worked for over 10 years in the non-profit field. She was a Program Officer in Scholarly Communications at The Mellon Foundation, funding non-profit publishers, libraries, preservation projects, and digital resources. She is currently Executive Director of the Society for Classical Studies.

Shawn Graham trained in Roman archaeology, but styles himself a digital archaeologist these days; he plays with digital techs, thinking about the way these break and what the breakages imply for archaeological thought. He is currently an associate professor in the Department of History at Carleton University in Ottawa, Canada

Sebastian Heath is an active Mediterranean archaeologist with long experience applying digital technologies to the study of the Ancient World. His research interests include Linked Open Data, Graph Databases, and 3D modeling. He is currently on the faculty at New York University's Institute for the Study of the Ancient World.

Eric Poehler is an Associate Professor at the University of Massachusetts Amherst, specializing in Roman Archaeology, Pompeii, and ancient infrastructure. He is currently directing or co-directing several digital and fieldwork projects including the Pompeii Bibliography and Mapping Project, the Pompeii Artistic Landscape Project (with Sebastian Heath), the Tharros Archaeological Research Project (with Steven Ellis), and the Pirene Valley Project at Corinth (with Betsey Robinson).

David M. Ratzan is Head of the ISAW Library at the Institute for the Study of the Ancient World at New York University. His main research and teaching interests are the social, economic, and administrative history of Greco-Roman antiquity, particularly as documented by the papyri that survive from Egypt. He has published on a wide array of topics related to Roman Egypt, including papyrology, numismatics, ancient law, and magic.

Lisl Walsh is an Associate Professor of Greek, Latin, and Ancient Mediterranean Studies at Beloit College in Wisconsin. Her academic interests include Classics pedagogy, disability studies, and materialist feminism, and her current research project explores how Senecan tragedy interfaces with Julio-Claudian Rome.

www.ingramcontent.com/pod-product-compliance
Lightning Source LLC
Chambersburg PA
CBHW031443040426
42444CB00007B/942